Essential Practical NMR for Organic Chemistry

Essential Practical NMR for Organic Chemistry

S.A. RICHARDS
AND
J.C. HOLLERTON

Second Edition

WILEY

Registered Office(s)
John Wiley & Sons, Inc., 111 River Street, Hoboken, NJ 07030, USA
John Wiley & Sons Ltd, The Atrium, Southern Gate, Chichester, West Sussex, PO19 8SQ, UK

Editorial Office
The Atrium, Southern Gate, Chichester, West Sussex, PO19 8SQ, UK

For details of our global editorial offices, customer services, and more information about Wiley products visit us at www.wiley.com.

Wiley also publishes its books in a variety of electronic formats and by print-on-demand. Some content that appears in standard print versions of this book may not be available in other formats.

Library of Congress Cataloging-in-Publication Data
Names: Richards, S.A., author. | Hollerton, J.C. - author.
Title: Essential practical NMR for organic chemistry / S.A. Richards and J.C. Hollerton.
Other titles: Essential practical nuclear magnetic resonance for organic chemistry
Description: Second edition. | Hoboken, NJ : John Wiley & Sons Ltd., 2023. | Includes index.
Identifiers: LCCN 2022035662 (print) | LCCN 2022035663 (ebook) | ISBN 9781119844808 (hardback) | ISBN 9781119844815 (epdf) | ISBN 9781119844822 (epub)
Subjects: LCSH: Proton magnetic resonance spectroscopy. | Nuclear magnetic resonance spectroscopy.
Classification: LCC QD96.P7 E87 2023 (print) | LCC QD96.P7 (ebook) | DDC 543/.66--dc23/eng/20221212
LC record available at https://lccn.loc.gov/2022035662
LC ebook record available at https://lccn.loc.gov/2022035663

Cover image: © Yuichiro Chino/Getty Images
Cover design by Wiley

Set in 10.5/12.5pt TimesNewRoman by Integra Software Services Pvt. Ltd, Pondicherry, India

Printed and bound by CPI Group (UK) Ltd, Croydon CR0 4YY

C9781119844808_231222

We would like to dedicate this book to our families and our NMR colleagues past and present.

Contents

Preface

This second edition of *Essential Practical NMR for Organic Chemistry* is an updated and improved version of the first edition which was a follow-up to the original *Laboratory Guide to Proton NMR Spectroscopy* (Blackwell Scientific Publications, 1988). It follows the same informal approach and is hopefully fun to read as well as a useful guide. While still concentrating on proton NMR, it includes 2-D approaches and some heteronuclear examples (specifically ^{13}C, ^{15}N and ^{19}F). This new edition now contains a comprehensive chapter on ^{15}N which we have found increasingly important in the last decade. The greater coverage is devoted to the techniques that you will be likely to make most use of.

The book is here to help you select the right experiment to solve your problem and to then interpret the results correctly. NMR is a funny beast – it throws up surprises no matter how long you have been doing it (at this point, it should be noted that the authors have more than 80 years of NMR experience between them and we still get surprises now and then!).

The strength of NMR, particularly in the small organic molecule area, is that it is very information rich but ironically, this very high density of information can itself create problems for the less experienced practitioner. Information overload can be a problem and we hope to redress this by advocating an ordered approach to handling NMR data. There are huge subtleties in looking at this data; chemical shifts, splitting patterns, integrals, linewidths all have an existence due to physical molecular processes and they each tell a story about the atoms in the molecule. There is a *reason* for *everything* that you observe in a spectrum and the better your understanding of spectroscopic principles, the greater can be your confidence in your interpretation of the data in front of you.

So, who is this book aimed at? Well, it contains useful information for anyone involved in using NMR as a tool for solving structural problems. It is particularly useful for chemists who have to run and look at their own NMR spectra and also for people who have been working in small molecule NMR for a relatively short time (less than 20 years, say;-)… It is focused on small organic molecule work (molecular weight less than 1500, commonly about 300). Ultimately, the book is pragmatic – we discuss cost-effective experiments to solve chemical structure problems as quickly as possible. It deals with some of the unglamorous bits, like making up your sample. These are necessary if dull. It also looks at the more challenging aspects of NMR.

While the book touches on some aspects of NMR theory, the main focus of the text is firmly rooted in data acquisition, problem-solving strategy and interpretation. If you find yourself wanting to know more about aspects of theory, we suggest the excellent, *High-Resolution NMR Techniques in Organic Chemistry* by Timothy D. W. Claridge (Elsevier, ISBN-13: 978–0-08–054818-0) as an approachable next step before delving into the even more theoretical works. Another really good source is Joseph P. Hornak's 'The Basics of NMR' website (you can find it by putting 'hornak nmr' into your favourite search engine). While writing these chapters, we have often fought with the problem of statements that are partially true and debated whether to insert a qualifier. To get across the fundamental ideas we have tried to minimise the disclaimers and qualifiers. This aids clarity, but be aware, almost everything is more complicated than it first appears!

Forty years in NMR has been fun. The amazing thing is that it is still fun…and challenging… and stimulating even now!

Please note that all spectra included in this book were acquired at 400 MHz unless otherwise stated.

1

Getting Started

1.1 The Technique

This book is not really intended to give an in-depth education on all aspects of the NMR effect (there are numerous excellent texts if you want more information), but we will try to deal with some of the more pertinent ones.

The first thing to understand about NMR is just how insensitive it is compared with many other analytical techniques. This is because of the origin of the NMR signal itself.

The NMR signal arises from a quantum mechanical property of nuclei called 'spin'. In the text here, we will use the example of the hydrogen nucleus (proton) as this is the nucleus that we will be dealing with mostly. Protons have a 'Spin Quantum Number' of ½. In this case, when they are placed in a magnetic field there are two possible spin states that the nucleus can adopt and there is an energy difference between them (Figure 1.1).

The energy difference between these levels is very small, which means that the population difference is also small. The NMR signal arises from this population difference and hence the

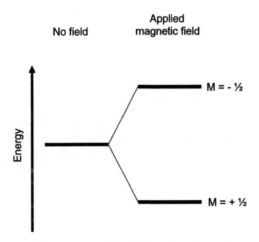

Figure 1.1   Energy levels of spin ½ nucleus.

Essential Practical NMR for Organic Chemistry, Second Edition. S.A. Richards and J.C. Hollerton.
© 2023 John Wiley & Sons Ltd. Published 2023 by John Wiley & Sons Ltd.

signal is also very small. There are several factors which influence the population difference and these include the nature of the nucleus (its 'Gyromagnetic ratio') and the strength of the magnetic field that they are placed in. The equation that relates these factors (and almost the only one in this book) is shown here:

$$\Delta E = \frac{\gamma h B}{2\pi}$$

E = Energy

γ = Gyromagnetic ratio

h = Planck's constant

B = Magnetic field strength

Because the sensitivity of the technique goes up with magnetic field, there has been a drive to increase the strength of the magnets to improve sensitivity.

Unfortunately, this improvement has been linear since the first NMR magnets (with a few kinks here and there). This means that in percentage terms, the benefits have become smaller as development has continued. But sensitivity has not been the only factor driving the search for more powerful magnets. You also benefit from stretching your spectrum and reducing overlap of signals when you go to higher fields. Also, when you examine all the factors involved in signal to noise, the dependence on field is to the power of 3/2 so we actually gain more signal than a linear relationship. Even so, moving from 800 to 900 MHz only gets you a 10% increase in signal to noise whereas the cost difference is considerably more than that!

In order to get a signal from a nucleus, we have to change the populations of each spin state. We do this by using tuned radio frequency to excite the nuclei into their higher energy state. We can then either monitor the absorption of the energy that we are putting in or monitor the energy coming out when nuclei return to their low energy state.

The strength of the NMR magnet is normally described by the frequency at which protons resonate in it – the more powerful the magnet, the higher the frequency. The earliest commercial NMR instruments operated at 40 megahertz (MHz) (megacycles in those days) whereas modern NMR magnets are typically ten times as powerful and the most potent (and expensive!) machines available can operate at fields of over 1 GHz.

1.2   Instrumentation

So far, we have shown where the signal comes from, but how do we measure it? There are two main technologies: Continuous Wave (CW) and Pulsed Fourier Transform (FT). CW is the technology used in older systems and is becoming hard to find these days (we only include it for the sake of historical context and because it is perhaps the easier technology to understand). FT systems offer many advantages over CW and they are used for almost all modern instruments.

1.2.1   CW Systems

These systems work by placing a sample between the pole pieces of a magnet (electromagnet or permanent), surrounded by a coil of wire. Frequency-swept radio frequency (RF) is fed into the wire, or alternatively, the magnet may have extra coils built onto the pole pieces which can be used to sweep the field with a fixed frequency. When the combination of field and frequency match the resonant frequency of a nucleus, radiofrequency is absorbed by the nuclei

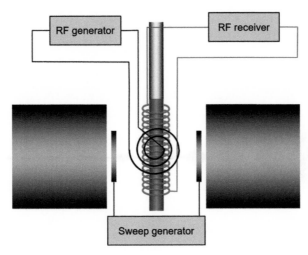

Figure 1.2   Schematic of a CW NMR spectrometer.

and re-emitted and captured by a receiver coil perpendicular to the transmitter coil (Figure 1.2). This emission intensity is then plotted against frequency. The whole process of acquiring a spectrum using a CW instrument takes typically about 5 minutes. Each signal is brought to resonance sequentially and the process cannot be rushed!

1.2.2   FT Systems

Most spectrometers used for the work we do today are Fourier transform (FT) systems. More correctly, they are pulsed Fourier transform systems. Unlike CW systems, the sample is exposed to a powerful polychromatic pulse of radio frequency. This pulse is very short and so contains a spread of frequencies (this is basic Fourier theory and is covered in Chapter 2). The result is that all of the signals of interest are excited simultaneously (unlike CW where they are excited sequentially) and we can acquire the whole spectrum in one go. This gives us an advantage (known as the Felgett advantage) in that we can acquire a spectrum in a few seconds as opposed to several minutes with a CW instrument. Also, because we are storing all this data in a computer, we can perform the same experiment on the sample repeatedly and add the results together. The number of experiments is called the number of scans (or transients, depending on your spectrometer vendor). Because the signal is coherent and the noise is random, we improve our signal to noise with each transient that we add. Unfortunately, this is not a linear improvement because the noise also builds up albeit at a slower rate (due to its lack of coherence). The real signal to noise increase is proportional to the square root of the number of scans (more on this later).

So if the whole spectrum is acquired in one go, why can't we pulse really quickly and get thousands of transients? The answer is that we have to wait for the nuclei to lose their energy to the surroundings. This takes a finite time and for most protons is just a few seconds (under the conditions that we acquire the data). So, in reality we can acquire a new transient every 3 or 4 seconds.

After the pulse, we wait for a short while (typically a few microseconds), to let that powerful pulse ebb away, and then start to acquire the radio frequency signals emitted from the sample. This exhibits itself as a number of decaying cosine waves. We term this pattern the 'free induction decay' or FID (Figure 1.3).

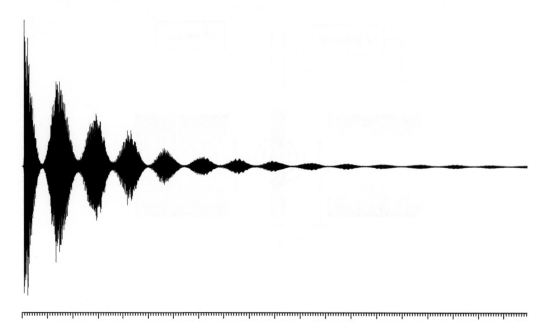

0.1 0.2 0.3 0.4 0.5 0.6 0.7 0.8 0.9 1.0 1.1 1.2 1.3 1.4 1.5 1.6 1.7 1.8   **sec**

Figure 1.3   A free induction decay (FID).

Obviously this is a little difficult to interpret, although with experience you can train yourself to extract all the frequencies by eye… (only kidding!). The FID is a 'time-domain' display but what humans really need is a 'frequency-domain' display (with peaks rather than cosines). To bring about this magic, we make use of the work of Jean Baptiste Fourier (1768–1830) who was able to relate time-domain to frequency-domain data. These days, there are super-fast algorithms (the Cooley-Tukey algorithm, for example) to do this and it all happens at the press of a button. It is worth knowing a little about this relationship as we will see later when we discuss some of the tricks that can be used to extract more information from the spectrum.

There are many other advantages with pulsed FT systems in that we can create trains of pulses to make the nuclei perform 'dances' which allow them to reveal more information about their environment. Ray Freeman coined the rather nice term 'Spin Choreography' to describe the design of pulse sequences. If you were fortunate enough to be able to attend one of his lectures, Ray Freeman had the gift of making this complex, mathematical subject easy to understand. Ray's NMR textbooks are a great source of information if you want to understand more. In particular, his book, *Spin Choreography Basic Steps in High Resolution NMR* (Oxford University Press, ISBN 0–19-850481–0).

Because we now operate with much stronger magnets than in the old CW days, the way that we generate the magnetic field has changed. Permanent magnets are not strong enough for fields above 90 MHz and conventional electromagnets would consume far too much electricity to make them viable (they would also be huge in order to keep the coil resistance low and need cooling to combat the heating effect of the current flowing through the magnet coils). The advent of superconducting wire made higher fields possible.

(The discovery of superconduction was made at Leiden University, by Heike Kamerlingh Onnes back in 1911 while experimenting with the electrical resistance of mercury, cooled to liquid helium temperature. His efforts were recognised with a Nobel Prize for Physics in 1913 and much later, a crater on the dark side of the moon was named after him. The phenomenon was to pave the way to the development of the very powerful superconducting magnets that we have today…)

Superconducting wire has no resistance when it is cooled below a critical temperature. For the wire used in most NMR magnets, this critical temperature is slightly above the boiling point of liquid helium (which boils at just over 4 Kelvin or about −269°C). (It should be noted that new superconducting materials are being investigated all the time. At the time of writing, some ceramic superconductors can become superconducting at close to liquid nitrogen temperatures although these can be tricky to make into coils.) When a superconducting magnet is energised, current is passed into the coil below its critical temperature. The current continues to flow undiminished, as long as the coil is kept below the critical temperature. To this end, the magnet coils are immersed in a Dewar of liquid helium. Because helium is expensive (believe it or not, it comes from holes in the ground) we try to minimise the amount that is lost through boil-off, so the liquid helium Dewar is surrounded by a vacuum and then a liquid nitrogen Dewar (temperature −196°C). A schematic diagram of a superconducting magnet is shown in Figure 1.4. Obviously, our sample can't be at −269°C (it wouldn't be very liquid at that temperature) so there has to be very good insulation between the magnet coils and the sample measurement area.

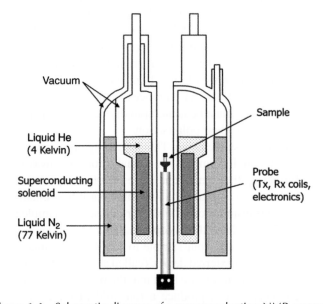

Figure 1.4   Schematic diagram of a superconducting NMR magnet.

1.2.3   Probes

In the centre (room temperature) part of the magnet we also need to get the radiofrequency coils and some of the tuning circuits close to the sample. These are normally housed in an aluminium cylinder with some electrical connectors and this is referred to as the 'probe'. The NMR tube containing the sample is lowered into the centre of the magnet using an air lift. The tube itself is long and thin (often 5 mm outside diameter) and designed to optimise the filling of the receive coil in the probe. We would call such a probe a '5 mm probe' (for obvious reasons!). It is also possible to get probes with different diameters and the choice of probe is made based on the typical sample requirements. At the time of writing, common probes go from 1 mm outside diameter (pretty thin!) to 10 mm although there are some other special sizes made.

Probes are designed to look at a specific nucleus or groups of nuclei. A simple probe would be a proton–carbon dual probe. This would have two sets of coils and tuning circuits, one for carbon the other for proton. Additionally, there would be a third circuit to monitor deuterium. The reason for having a deuterium signal is that we can use this signal to 'lock' the spectrometer frequency so that any drift by the magnet will be compensated by monitoring the deuterium resonance (more on this later).

There is a vast array of probes available to do many specialist jobs but for the work that we will discuss in this book, a proton–carbon dual probe would perform most of the experiments (although having a four-nucleus probe is better as this would allow other common nuclei such as fluorine or phosphorus or nitrogen to be observed).

Probes can have one of two geometries. They can be 'normal' geometry, in which case the non-proton nucleus coils would be closest to the sample or 'inverse' geometry (the inverse of normal!). We mention this because it will have an impact on the sensitivity of the probe for acquiring proton data (inverse is more sensitive than normal). Most of the time this shouldn't matter unless you are really stuck for sample in which case it is a bigger deal...

Because sensitivity has always been the Achilles Heel of NMR, we are always looking for ways to improve it. If we want to improve sensitivity, then we need to either increase the signal or decrease the noise (or both). There are approaches that can effectively pump nuclei into the higher energy state (e.g. Dynamic Nuclear Polarisation). These approaches have been shown to massively increase signal although not uniformly across a molecule. They are showing a lot of promise in solids NMR and certain niche areas but are not commonly used for small molecule work because of the lack of uniformity of signal enhancement. There is also a potential for sample degradation as the Nuclear Polarisation process happens at very low temperatures outside the main magnet and samples must be heated rapidly before entering the NMR magnet so that they are liquid and haven't lost their polarisation.

The other approach is to decrease the noise. This is nothing new; low-noise electronics have been improved over the years, but it is impossible to make noise go away once it has been generated. This pushes us further up the signal generation path to the probe and preamplifier. The idea of cooling the probe has been around for a long time but it has remained a technical challenge. How do we keep cold transmit/receive coils very close to a sample at 30°C? Imagine if those cold coils are at −250°C! Enter the cryoprobe. This is a miracle of engineering with modern versions allowing you to heat your sample to 120°C whilst maintaining your coils at −250°C. The coils are cooled by an external device (called a cryoplatform) which passes helium at ~20 K over the coils. The preamplifiers are built into the probe body and these too are cooled. All of this incredible engineering can produce a four-fold increase in signal-to-noise. This means that you can get away with a quarter of the amount of material to get the same signal in the same time. Perhaps more importantly, due to the relationship between signal to noise and time, you can acquire the same data on the same weight of sample in a sixteenth of the time!

Cryoprobes are not without their disadvantages though. They cost a lot of money to buy and they also cost a lot to run. Making things very cold requires a lot of energy – around 12 kW constantly!

1.2.4 Shims

Because we are looking at tiny differences in frequency as a proportion of the main frequency (e.g. 0.5 Hz in a 600 MHz field), the magnetic field must be homogeneous to the same level. Building probes requires extreme precision but, even with this, it is impossible to make the field homogeneous to the necessary level. The magnet is designed with a number of superconducting coils (shims) that attempt to get the field as good as possible but even this is not enough. To get the last precision on the field, there are also some room-temperature shims which live between the probe body and the inner magnet wall. These are tweaked for every sample to get the best possible signal. More about this in Chapter 3.

1.3 Origin of the Chemical Shift

Early NMR experiments were expected to show that a single nucleus would absorb radio frequency energy at a discrete frequency and give a single line. Experimenters were a little disconcerted to find, instead, some 'fine structure' on the lines and when examined closely, in some cases, lots of lines spread over a frequency range. In the case of proton observation, this was due to the influence of surrounding nuclei shielding and deshielding the close nuclei from the magnetic field. The observation of this phenomenon gave rise to the term 'chemical shift', first observed by Fuchun Yu and Warren Proctor in 1950. There were some who thought this to be a nuisance, but it turned out to be the effect that makes NMR such a powerful tool in solving structural problems.

There are many factors that influence the chemical shift of an NMR signal. Some are 'through bond' effects such as the electronegativity of the surrounding atoms. These are the most predictable effects and there are many software packages around which do a good job of making through bond chemical shift predictions. Other factors are 'through space' and these include electric and magnetic field effects. These are much harder things to predict as they are dependent on the average solution conformation of the molecule of interest.

In order to have a reliable measurement of chemical shift, we need to have a reference for the value. In proton NMR this is normally referenced to tetramethyl silane (TMS) which is notionally given the chemical shift of zero. A spectrum of TMS would look like this (Spectrum 1.1).

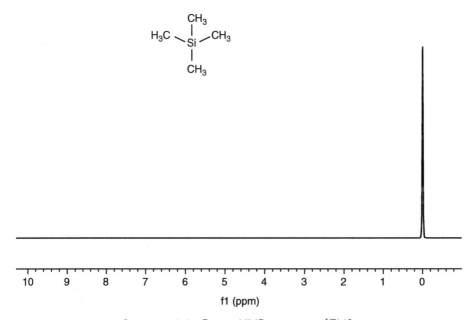

Spectrum 1.1   Proton NMR spectrum of TMS.

You will notice that the spectrum runs 'backwards' compared with most techniques (i.e. '0' is at the right of the graph). This is because the silicon in TMS shields the protons from the magnetic field. Most other signals will come to the left of TMS. For some years, there was a debate about this and there were two different scales in operation. The scale shown here is the now accepted one and is called 'δ'. The older scale (which you may still encounter in old literature) is called 'τ' and it references TMS at 10 so you need a little mental agility to make the translation between the two scales. The scale itself is quoted in ppm (parts per million). It is actually a frequency scale, but if we quoted the frequency, the chemical shift would be dependent on the magnetic field (a 400 MHz spectrometer would give different chemical shifts to a 300 MHz spectrometer). To get around this, the chemical shift is quoted as a ratio compared with the main magnet field and is quoted in ppm.

Finally, we have an issue with how we describe relative chemical shifts. Traditionally (from CW NMR days) we describe them as 'upfield' (to lower delta) and 'downfield' (to higher delta). This is not strictly correct in a pulsed FT instrument (because the field remains static) but the terminology continues to be used. We still use these terms in this book as the alternatives are a bit cumbersome.

1.4   Origin of 'Splitting'

So far, we have seen where NMR signals come from, and touched on why different groups of protons have different chemical shifts. In addition to the dispersion of lines due to chemical shift, if you look closely the individual lines may be split further. If we take the example of ethanol, this becomes obvious (Spectrum 1.2). We now have to understand why some signals appear as multiple lines rather than just singlets. Protons, that are chemically and magnetically distinct from each other, interact magnetically if they are close enough to do so by the process known as 'spin–spin coupling'. 'Close enough' in this context means 'separated by two, three, or

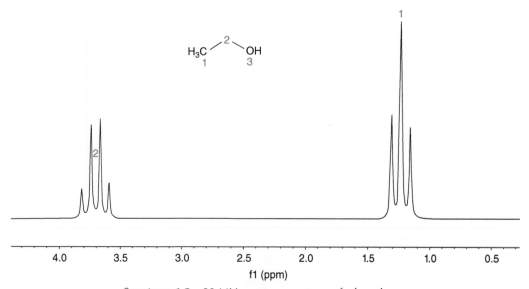

Spectrum 1.2   90 MHz proton spectrum of ethanol.

occasionally four bonds'. Let us consider an isolated ethyl group such as found in ethanol. (We will assume no coupling from the –OH proton for the moment.)

On examining Spectrum 1.2, you will notice that the –CH$_2$– protons appear as a four-line quartet, while the –CH$_3$ protons give a three-line triplet. Furthermore, the relative intensities of the lines of the quartet are in the ratio, 1:3:3:1, while the triplet lines are in the ratio 1:2:1.

We'll consider the methyl triplet first. While the signal is undergoing irradiation, the methylene protons are, of course, aligned either with, or against the external magnetic field as discussed earlier. Note that as far as spin–spin coupling is concerned, we may consider the two states to be equally populated. If we call the methylene protons H$_A$ and H$_B$, then at any time, H$_A$ and H$_B$ may be aligned with the external magnetic field, or against it. Alternatively, H$_A$ may be aligned with the field, while H$_B$ is aligned against it, or vice versa, the two arrangements being identical as far as the methyl protons are concerned.

So, the methyl protons experience different magnetic fields depending on the orientation of the methylene protons. The statistical probability of one proton being aligned with and one against the magnetic field is twice as great as the probability of both being aligned either with, or against the field. This explains why the relative intensity of the methyl lines is 1:2:1. Spin–spin coupling is always a reciprocal process – if protons '*x*' couple to protons '*y*', then protons '*y*' must couple to '*x*'. The possible alignments of the methyl protons (H$_C$, H$_D$ and H$_E$) relative to the methylene protons are also shown in Spectrum 1.2. Think about the orientations of protons responsible for multiplet systems as we meet them later on.

There are two other important consequences of spin–spin coupling. Firstly, *n* equivalent protons will split another signal into *n* + 1 lines (hence three methyl protons split a methylene CH$_2$ into 3 + 1 = 4 lines). Secondly, the relative sizes of peaks of a coupled multiplet can be calculated from Pascal's triangle (Figure 1.5):

Splitting pattern								Number of adjacent protons	Description
				1				0	singlet
			1		1			1	doublet
		1		2		1		2	triplet
	1		3		3		1	3	quartet
1		4		6		4		1	quintet
1	5		10		10		5	1	5	sextet
1	6	15		20		15	6	1	6	septet

Figure 1.5   Pascal's triangle.

We have often found that students have a touching but misplaced faith in Mr. Pascal and his triangle and this can lead to no end of angst and confusion! It is very important to note that you will only come across this symmetrical distribution of intensities within a multiplet when the signals coupling to each other ALL SHARE THE SAME COUPLING CONSTANT – as soon as a molecule gains a chiral centre and couplings from neighbouring protons cease to be equivalent, Pascal's triangle ceases to have any value in predicting the appearance of multiplets. Also, coupled signals must be well separated in order to approximately adhere to Pascal's distribution. This obviously begs the question: 'How well separated?' Well, this is a tricky question to answer. It is not possible to put an absolute figure on it because the further away the coupling signals are from each other in the spectrum, the better will be the concord between the theoretical distribution of intensity and the actual one. We will talk about this problem again later. Well-separated coupled signals give rise to 'first-order' spectra and poorly separated ones give rise to 'non-first-order' spectra. We'll see examples of both types in due course.

An important point to remember is that the signal intensity for the proton is shared between these lines. Pascal's triangle gives the *relative* intensities of the lines in the multiplet, not the absolute ones (otherwise our integrals wouldn't make sense as seen in the next section). This means that, for the same number of protons, the peaks of a doublet will be half the intensity of a singlet. For a triplet, the outer peaks would have ¼ of the intensity of a singlet and the central peak would have half the intensity of the singlet. If you have very little material for your NMR and your signal to noise is compromised, you may struggle to see complex multiplets as their intensity will be shared over many peaks and this may drop them into the noise. Perhaps a better way of representing the triangle is to show intensities. Figure 1.6 shows the intensities for multiplets up to quartets. At the level of a quartet, the outer lines have an eighth of the intensity of a singlet. The outer lines of a septet are only 1/64th of the size of a singlet so you will often struggle to see these lines unless you have a significant amount of material (or a very sensitive instrument).

The separations between the lines of doublets, triplets and multiplets are very important parameters and referred to as 'coupling constants', though the term is not strictly accurate. 'Measured splittings' would be a better description, since true coupling constants can only be measured in totally first-order spectra (which implies infinite separation between complex signals), which never exist in practice. However, the differences between true coupling constants and measured splittings are so small for reasonably first-order spectra, that we shall overlook any discrepancies which are vanishingly small anyway.

We measure coupling constants in Hz, since if we measured them in fractions of a ppm, they would not be constant, but would vary with the magnetic field strength of the spectrometer used. This would obviously be most inconvenient! Note that 1 ppm = 250 Hz on a 250 MHz spectrometer and 400 Hz on a 400 MHz spectrometer, etc.

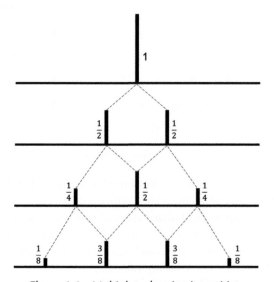

Figure 1.6   Multiplets showing intensities.

1.5 Integration

The area of each signal is proportional to the number of nuclei at that chemical shift. If we look at the previous example, the signal for the methyl group in ethanol should have an area with the ratio of 3:2 compared with the methylene signal. When we plot proton NMR data, we usually also plot the integral as well. This will show us the relative areas under the curves. Spectrum 1.3 shows the spectrum of ethanol with integrals.

Often, the integrals are broken up to maximise their size on the display and make them easier to measure. Integrals can be tricky to measure exactly, especially if the signal to noise of the spectrum is low or if the baseline is distorted. Overlapping signals also make it difficult to integrate accurately and so other tools are available to perform peak fitting and use the peak parameters to back-calculate the integrals. If you want approximate integrals, then this is fine. On the other hand, if you are trying to integrate very accurately (e.g. to calculate purity values) then you need to take a lot of factors into consideration (see Chapter 13 for more details).

Note that the height of a peak is not equivalent to its integral! If you have singlets, then you could probably approximate their height to their integral but only if you know that their linewidths are the same. The only true measurement for number of protons is using the integral.

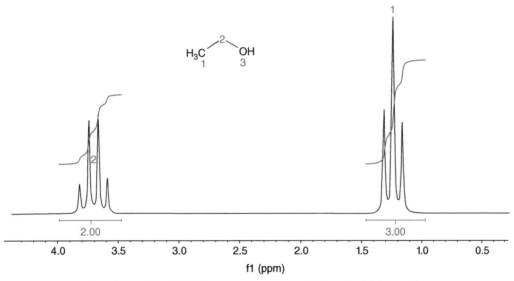

Spectrum 1.3   90 MHz proton spectrum of ethanol with integrals.

The area of each signal is proportional to the number of nuclei at that chemical shift. If we look at the previous example, the signal for the methyl group in ethanol should have an area with the ratio of 3:2 compared with the methylene signal. When we plot proton NMR data, we usually also plot the integral as well. This will show us the relative areas under the curve. Spectrum 1.3 shows the spectrum of ethanol with integrals.

Often, the integrals are broken up to maximise their size on the display and make them easier to measure. Integrals can be tricky to measure exactly, especially if the signal to noise of the spectrum is low or if the baseline is distorted. Overlapping signals also make it difficult to integrate accurately and so other tools are available to perform peak fitting and also the peak parameters to back-calculate the integrals. If you want approximate integrals, then this is fine. On the other hand, if you are trying to integrate very accurately (e.g. to calculate purity values) then you need to take a lot of factors into consideration (see Chapter 13 for more details).

Note that the height of a peak is not equivalent to its integral. If you have singlets, then you could roughly approximate their height to their integral but only if you know that their linewidths are the same. The only true measurement for number of protons is using the integral.

Spectrum 1.3 Proton NMR spectrum of ethanol with integrals.

f1 (ppm)

2

Preparing the Sample

While sample preparation may not be the most interesting aspect of NMR spectroscopy, it is nonetheless *extremely* important as it will have a huge bearing on the quality of the data obtained and therefore on your ability to make logical deductions about your compounds. This is particularly true when acquiring the most straightforward 1-D proton spectra. The most typical manifestation of sub-standard sample preparation is poor lineshape. It is worth remembering that in terms of 1-D proton NMR, 'the devil' can be very much 'in the detail'. 'Detail', in this context, means 'fine structure' and fine structure is always the first casualty of poor sample preparation.

The reason for this can best be appreciated by considering just how small the differences in chemical shifts of signals really are – and indeed, just how small (but significant!) a long-range coupling can be. Consider for example, a 3–7 coupling in an indole.

Being able to see this coupling is reassuring in that it ties the 3 and 7 protons together for us. It might seem a trifling matter, but observing it, even if it appears only as a slight but definite broadening, helps underpin the credentials of the molecule because we know it should be there. Such a 5-bond coupling will be small – comparable in fact with the natural line width of a typical NMR signal. Let's say we are looking for a coupling of around 1 Hz, for the sake of argument. 1 Hz, on a 400 MHz spectrometer corresponds to only 1/400 of a part per million of the applied magnetic field (since 1 ppm = 400 Hz in a 400 MHz spectrometer). So in order to observe such a splitting, we will need resolution of better than 0.5 Hz which corresponds to 1 part in $10^6/(0.5/400)$ or ideally, better than 1 part in 10^9! To achieve such resolution requires corresponding levels of magnetic field homogeneity through your sample but this can only be

Essential Practical NMR for Organic Chemistry, Second Edition. S.A. Richards and J.C. Hollerton.
© 2023 John Wiley & Sons Ltd. Published 2023 by John Wiley & Sons Ltd.

achieved in extremely clean solutions of sufficient depth. We will be dealing with this issue in detail later on. In real terms, establishing first-class magnetic field homogeneity means that molecules of your compound will experience exactly the same field no matter where they are in the NMR tube – therefore, they will all resonate in unison – rather than in a fragmented fashion. Any factor which adversely effects field homogeneity will have a corresponding deleterious effect on line shape. We will see this more clearly later.

2.1   How Much Sample Do I Need?

This section might be alternatively titled, 'How long is a piece of string?' There is no simple answer to this question which we have been asked many, many times. What you need in solution is sufficient material to produce a spectrum of adequate signal to noise to yield the required information but this is no real answer as it will vary with numerous factors. How powerful is the magnet of the spectrometer you are using? What type of probe is installed in it? What nucleus are you observing? What type of NMR acquisition are you attempting? How pure is your sample? What is the molecular weight of your sample? Is it a single compound or is it a mixture of diastereoisomers? These are just some of the relevant questions that you should consider.

And there are others. If you are using a walk-up system, there will probably be some general guidelines posted on it. Assume that these are useful and adhere to them as far as possible. They will be, by their very nature, no more than a guide, as every sample is unique in terms of its molecular weight and distribution of signal intensity. Also, a walk-up system is likely to be limited in terms of how much time (and therefore how many scans) it can spend on each sample.

If you are fortunate enough to be 'driving' the spectrometer yourself, you can of course compensate for lack of sample by increasing the number of scans you acquire on your sample – but this is not a license to use vanishingly small amounts. It is worth remembering that in order to double the signal/noise ratio, you have to acquire four times the number of scans. Think about it. If your sample is still giving an unacceptably noisy spectrum after 5 minutes of acquisition, how long will you have to leave it acquiring in order for the signal/noise (S/N) to become acceptable? Doubling the S/N is likely to do little. If you improve it by a factor of four (probably a worthwhile improvement) you will have to acquire for an hour and 20 minutes (16×5 minutes)! The law of diminishing returns operates here and makes its presence felt very quickly indeed.

All that having been said, we will attempt to draw up a few rough guidelines below…

If you are unfortunate enough to be struggling away with some old continuous-wave museum piece, then in all probability, you will only be looking at proton spectra. Even though the proton is the most sensitive of all nuclei, you will still be needing *at least* 15 mg of compound, assuming a molecular weight of about 300 (if it's a higher molecular weight, you will need more material, lower and you may get away with a little less).

It's more likely these days that you will be using a 250 or 400 MHz Fourier transform instrument with multi-nuclei capability. If such an instrument is operating in 'walk-up' mode so that it can acquire in excess of 60 samples in a working day, then it will probably be limited to about 32 scans per sample (a handy number – traditionally, the number of scans acquired has always been a multiple of eight but we won't go into the reasons here. If you want more information, take a look at the term 'phase cycling' in one of the excellent texts available on the more technical aspects of NMR). This means that for straightforward 1-D proton acquisition, you will need about 3 mg of compound as above, though you may get away with as little as 1 mg with a longer acquisition time, assuming a typical 5 mm probe. The same 3 mg solution (sticking with the approx. 300 mol wt. throughout) would also get you a reasonable Fluorine spectrum, if available, since the ^{19}F nucleus is a 100% abundant and is, therefore, a relatively sensitive nucleus.

If you are looking for a ^{13}C spectrum, then you will probably find that they will only be available overnight. This is because the ^{13}C nucleus is extremely insensitive and acquisition will take hours rather than minutes (only 1.1% natural abundance and relatively low gyromagnetic ratio – see Glossary). While the signal to noise available for ^{13}C spectra will be highly dependent on the type of probe used (i.e. 'normal' geometry or 'inverse' geometry – see Glossary), about 20 mg of compound will be needed for a typical acquisition, which will probably entail about 3200 scans and run for about 2 hours. Even then, the signal/noise for the least sensitive quaternary carbons may well prove marginal. (Note that the inherently low sensitivity of the ^{13}C nucleus can to some extent be addressed by acquiring various inverse-detected 2-D data such as HMQC/HSQC and HMBC, all of which we will discuss later.)

Operating at 500 or 600 MHz and using a 3 mm probe should yield an approximate three-fold improvement in signal/noise which can be traded for a corresponding reduction in sample requirement.

Various technologies do exist to give still greater sensitivities – perhaps even an order of magnitude greater e.g. 'nano' probes, 1 mm probes and cryo-probes but they are currently unusual in a 'routine' NMR environment. These tools tend to be the preserve of the NMR specialist.

Table 2.1 offers a very rough guide to the amount of sample you need given all the previous provisos. Of course, if you are prepared to wait a long time and don't have a queue of people waiting to use the instrument, you can get away with less material. Generally, more is better (as long as the solution is not so gloopy that it broadens all the lines!).

Table 2.1   Amount of material required.

Field (MHz)	Comfortable amount of material required (mg)	
	^{1}H	^{13}C
90	20	Lots!
250	5	30
400	2	10
600	1	5
600 (cryo)	0.1	0.5

2.2   Solvent Selection

The first task when running any liquid-phase NMR experiment is the selection of a suitable solvent. Obvious though this sounds, there are a number of factors worth careful consideration before committing precious sample to solvent. A brief glance at any NMR solvents catalogue will illustrate that you can purchase deuterated versions of just about any solvent you can think of but we have found that there is little point in using exotic solvents when the vast majority of compounds can be dealt with using one of four or five basic solvents.

Your primary concern when selecting a solvent should be the **complete** dissolution of your sample. Again, this might seem an unduly trivial observation, but if your sample is not in solution, then it will remain 'invisible' to the spectrometer. Consider for a moment a hypothetical sample – a mixture of several components, only one of which being soluble in your chosen solvent. Under these circumstances, your spectrum may flatter you (your desired compound is preferentially soluble in solvent of choice), or alternatively, it may paint an unduly pessimistic view of your sample (one or more of the undesired components is preferentially soluble in solvent of choice). Either way, there are possibilities for being misled here so the primary objective in selecting a solvent should be the total dissolution of your sample. In general, we advise adhering to the simple old rule that 'like dissolves like'. In other words, if your sample is non-polar, then choose a non-polar solvent and vice versa.

2.2.1   Deutero Chloroform (CDCl$_3$)

This is a most useful NMR solvent. It can dissolve compounds of reasonably varying polarity, from non-polar to considerably polar, and the small residual CHCl$_3$ signal at 7.27 ppm seldom causes a problem. CDCl$_3$ can easily be removed by 'blowing-off' should recovery of the sample be necessary. Should a compound prove only sparingly soluble in this solvent, deutero dimethyl sulfoxide can be added drop by drop to increase the polarity of the solvent – but see cautionary notes below! This may be preferable to running in neat D$_6$-DMSO due to the disadvantages of D$_6$-DMSO outlined below. It should be noted that D$_6$-DMSO causes the residual CHCl$_3$ signal to move downfield to as low as 8.38 ppm, its position providing a rough guide to the amount of D$_6$-DMSO added. The main disadvantage of using a mixed solvent system is the difficulty of getting reproducible results, unless you take the trouble of measuring the quantities of each solvent used!

It should also be noted that CDCl$_3$ is best avoided for running spectra of salts, even if they are soluble in this solvent. This is because deutero chloroform is an 'aprotic' solvent that does not facilitate fast transfer of exchangeable protons. For this reason, spectra of salts run in this solvent are likely to be broad and indistinct as the spectrometer 'sees' two distinct species of compound in solution; one with a proton attached and another with it detached. As the process of inter-conversion between these two forms is slow on the NMR timescale (i.e. the time taken for the whole process of acquiring a single scan to be completed in), this results in averaging of the chemical shifts and consequent broadening of signals – particularly those near the site of protonation.

2.2.2   Deutero Dimethyl Sulfoxide (DMSO)

D$_6$-DMSO is undoubtedly very good at dissolving things. It can even dissolve relatively insoluble heterocyclic compounds and salts, but it does have its drawbacks. Firstly, it's relatively viscous, and this causes some degree of line-broadening. In cases of salts, where the acid is relatively weak (fumaric, oxalic, etc.), protonation of the basic centre may well be incomplete. Thus, salts of these weak acids may often look more like free-bases! It is also a relatively mild oxidising agent, and has been known to react with some compounds, particularly when warming the sample to aid dissolving, as is often required with this solvent.

Problems associated with restricted rotation (discussed later) also seem to be worse in D$_6$-DMSO, and being relatively non-volatile (it boils at 189°C, though some chemical decomposition occurs approaching this temperature so it is always distilled at reduced pressure), it is slightly harder to remove from samples, should recovery be required. This non-volatility, however, makes it the first choice for high-temperature work – it could be taken up to above 140°C in theory, though few NMR probes are capable of operating at such high temperatures.

At the other end of the temperature scale it is useless, freezing at 18.5°C. In fact, if the heating in your NMR lab is turned off at night, you may well find this solvent frozen in the morning during the cold winter months!

The worst problem with DMSO, however, is its affinity for water, (and for this reason, we recommend the use of sealed 0.75 ml ampoules wherever possible) which makes it almost impossible to keep dry, even if it's stored over molecular sieve. This means that bench D_6-DMSO invariably has a large water peak, which varies in shape and position, from sharp and small at around 3.5 ppm, to very large and broad at around 4.1 ppm in wetter samples. This water signal can be depressed and broadened further by acidic samples! This can be annoying as the signals of most interest to you may well be obscured by it. One way of combating this is to displace the water signal downfield by adding a few drops of D_2O, though this can also cause problems by bringing your sample crashing out of solution. If this happens, adding more D_6-DMSO to re-dissolve it is the best way forward. The residual CD_2HSOCD_3 signal occurs at 2.5 ppm, and is of characteristic appearance (caused by ^2H–^1H coupling). Note that the spin of deuterium is 1 which accounts for the complexity of the signal). See Spectrum 2.1. Even so-called 100% isotopic D_6-DMSO has a small residual signal so you can't totally negate the problem by using it – just lessen it.

Extreme care should be taken when handling DMSO solutions, as one of its other characteristics is its ability to absorb through the skin taking your sample with it! This can obviously be a source of extreme hazard. Wash off any accidental spillages with plenty of water – immediately! (This goes for all other solvents too.)

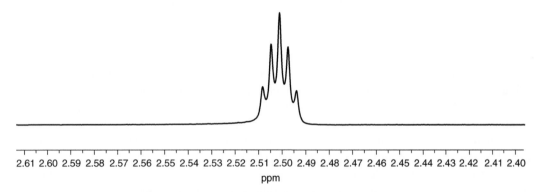

2.61 2.60 2.59 2.58 2.57 2.56 2.55 2.54 2.53 2.52 2.51 2.50 2.49 2.48 2.47 2.46 2.45 2.44 2.43 2.42 2.41 2.40

ppm

Spectrum 2.1   Residual solvent signal in DMSO.

2.2.3   Deutero Methanol (CD₃OD)

This is a very polar solvent, suitable for salts and extremely polar compounds. Like DMSO it has a very high affinity for water and is almost impossible to keep dry. Its water peak is sharper, and occurs more predictably at around 4.8 ppm. The residual CD_2HOD signal is of similar appearance to the D_6-DMSO residual signal, and is observed at 3.3 ppm.

Its main disadvantage is that it will exchange ionisable protons in your sample for deuterons, and hence they will be lost from the spectrum e.g. –OH, –NH, –SH and even –CONH₂, though these can often be relatively slow to exchange. Also, protons α to carbonyl groups may exchange through the keto-enol mechanism and aryl protons on aromatic rings containing two or more –OH groups by the similar keto-phenol mechanism. The importance of losing such information should not be underestimated. Solving a structural problem can often hinge on it!

2.2.4   Deutero Water (D₂O)

D_2O is even more polar than D_4-methanol and rather limited in its use for that reason – usually for salts only. Like deutero methanol, it exchanges all acidic protons readily and exhibits a strong HOD signal at about 4.9 ppm. Samples made up in D_2O often fail to dissolve cleanly and benefit from filtration through a tight cotton wool filter – cf. discussion of filtration later.

2.2.5   Deutero Benzene (C₆D₆)

D_6-Benzene is a rather specialised solvent and not normally used in 'routine' work. It is often added to $CDCl_3$ solutions, though it can of course be used neat, when it may reveal hidden couplings or signals by altering chemical shifts of your compound. It does this because it can form collision complexes with sample molecules by interactions of the pi electrons. This can bring about changes in the chemical shifts of the sample peaks because benzene is an anisotropic molecule i.e. it has non-uniform magnetic properties (shielding above and below the plane of the ring, and deshielding in the plane of the ring). This is really an extreme example of a solvent shift. Whenever you change the solvent, expect a change in the spectrum! C_6D_6 shows a residual C_6D_5H signal at 7.27 ppm. CAUTIONARY NOTE – Benzene is of course a well-known carcinogen and due care should be taken when handling it – particularly if used in combination with DMSO.

With these five solvents at your disposal, you will be equipped to deal with virtually any compound that comes your way but it might be worth briefly mentioning two others.

2.2.6   Carbon Tetrachloride (CCl₄)

This would be an ideal proton NMR solvent, (since it is aprotic and cheap) were it better at dissolving things! Its use is now very limited in practice to very non-polar compounds. Also, it lacks any deuterated signal that is required for locking modern Fourier transform spectrometers – (an external lock would be necessary making it inconvenient – see Section 2.3). Carbon tetrachloride is very hydrophobic, so any moisture in a sample dissolved in this solvent will yield a milky solution. This might impair homogeneity of the solution and therefore degrade resolution, so drying with anhydrous sodium sulfate can be a good idea. Carbon tetrachloride does have the advantage of being non-acidic, and so can be useful for certain acid-sensitive compounds. Take care when handling this solvent, as like benzene, it is known to be carcinogenic. Not recommended.

2.2.7   Trifluoroacetic Acid (CF₃COOH)

Something of a last resort this one! It seems to be capable of dissolving most things, but what sort of condition they're in afterwards is rather a matter of chance. It has been useful in the past for tackling extremely insoluble multicyclic heterocyclic compounds. If you have to use it, don't expect wonders. Spectra are sometimes broadened. It shows a very strong –COOH broad signal at about 11 ppm. Again, the lack of a deuterated signal in this solvent makes it less suitable for FT making an external lock necessary – see above. Not recommended unless no alternative available. Handle with care – it is extremely corrosive.

Well, that just about concludes our brief look at solvents. If you can't dissolve it in one of the common solvents, you've got problems. If in doubt, try a bit first, before committing your entire sample. Use non-deuterated solvents for solubility testing if possible, as they are much cheaper.

2.2.8 Using Mixed Solvents

While it is perfectly possible to use a mixed solvent system ($CDCl_3$/DMSO is always a popular example as chemists tend to opt for $CDCl_3$ out of habit or in the hope that it will dissolve their samples, only to find that solubility is not as good as expected), we advise against it, particularly if you are running your spectra on a 'walk-up' automated system. Remember that the spectrometer uses the deuterated signal for frequency locking and if it has more than one to choose from, things can go wrong and you might find yourself the proud owner of a spectrum that has been offset by several ppm as the spectrometer locks onto the D_6-DMSO signal and sets about its business in the belief that it has in fact locked onto $CDCl_3$! Furthermore, it is very difficult to reproduce exact solvent conditions if you are required to re-make a compound. Using a suitable single solvent will prevent these issues ever troubling you.

2.3 Spectrum Referencing (Proton NMR)

NMR spectroscopy differs from other forms of spectroscopy in many respects, one of which is the need for our measurement to be referenced to a known standard. For example, considering infra-red spectroscopy for a moment, if a carbonyl group stretches at 1730 cm^{-1}, then as long as we have a suitably calibrated spectrometer, we can measure this, confident in the knowledge that we are measuring an absolute value associated with that molecule.

In NMR spectroscopy, however, the chemical shift measurement we make takes place in an environment of our making that is both entirely artificial and arbitrary (i.e. the magnet). For this reason, it is essential to reference our measurements to a known standard so that we can all 'speak the same language', no matter what make or frequency of spectrometer we use.

The standard is usually added directly to the NMR solvent and is thus referred to as an 'internal' standard, though it is possible to insert a small tube containing standard in solvent into the bulk of the sample so that the standard does not come into direct contact with the sample. This would be referred to as an 'external' standard. We recommend an internal standard wherever possible for reasons of convenience and arguably superior shimming.

Apart from some very early work in the field which was performed using water as a standard (it would be difficult to imagine a worse reference standard as the water signal moves all over the place in response to changing pH!) the historical reference standard of choice has always been TMS (Tetra Methyl Silane), as mentioned earlier. TMS has much to recommend it as a standard. It is chemically very inert and is volatile (b.p. 26–28°C) and so can easily be removed from samples if required. Furthermore, only a tiny amount of it is needed as it gives a very strong twelve proton singlet in a region of the spectrum where other signals seldom occur.

TMS is not ideally suited for use in all solvents, however. As you can see from the structure, it is extremely non-polar and so tends to evaporate from the more polar solvents (D_6-DMSO and D_4-MeOD). For this reason, a more polar derivative of TMS (TSP – see below) is often used with these solvents.

3-(Trimethylsilyl) propionic-2,2,3,3-D_4 acid, sodium salt.

Note that the side chain is deuterated so that the only signal observed in the proton NMR spectrum is the trimethyl signal.

Deuterated solvents can be purchased with these standards already added if required and this would be our recommendation because so little standard is actually needed that it is very difficult to add little enough to a single sample without overdoing it. (An enormous standard peak, apart from looking amateurish, is to be avoided since it will limit signal/noise ratio as the spectrometer scales the build-up of signals according to the most intense peak in a spectrum.) Of course, TMS and TSP do not have *exactly* the same chemical shifts so, to be totally meticulous, you should really quote the standard you are using when recording data.

Of course, you don't have to use either of the above standards at all. In the case of samples run in deutero chloroform/methanol and dimethyl sulfoxide, it is perfectly acceptable, and arguably preferable, to reference your spectra to the residual solvent signal (e.g. CD_2HOH) which is unavoidable and always present in your spectrum (see below). These signals are perfectly solid in terms of their shifts (in pure solvent systems) though the same cannot be said for the residual HOD signal in D_2O and for this reason, we would advise adhering to TSP for all samples run in D_2O.

Solvent	Chemical shift of residual signal (ppm)
$CDCl_3$	7.27
CD_3SOCD_3	2.50
CD_3OD	3.30

We will discuss referencing issues with respect to other nuclei in later chapters.

2.4 Sample Preparation

Note that sample depth is important! When using a typical 5 mm probe, a sample depth of about 4 cm (approx. 0.6 ml) is necessary, though this varies slightly from instrument to instrument.

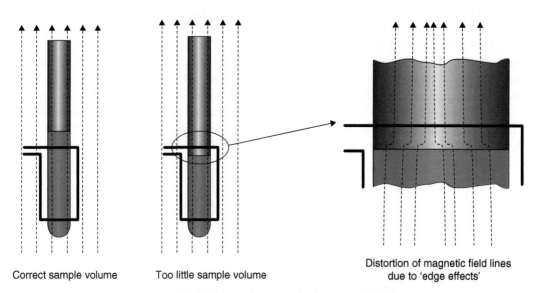

Correct sample volume   Too little sample volume   Distortion of magnetic field lines due to 'edge effects'

Figure 2.1 Sample depth and magnetic field homogeneity.

There should be guidance available to you in this respect on each individual spectrometer. If you try to get away with less than this, magnetic field homogeneity and therefore, shimming (see later), will be compromised as the transmitter and receiver coils in the probe must be covered to a sufficient depth to avoid the problems of 'edge effects'. See Figure 2.1.

Of course, there is no point in over-filling your NMR tubes. This *can* make shimming more difficult (but certainly not impossible as in the case of too low a sample depth) but more importantly, it merely wastes materials and gives rise to unduly dilute samples giving reduced signal/noise. Any sample outside the receiver coils does not contribute to the observed signal.

If your sample is reluctant to dissolve in the chosen solvent, avoid adding more solvent for the reasons outlined above. Instead, try warming the sample vial carefully on a hotplate or with a hairdryer – sometimes a bit of thermal agitation will be all that is required to assist the dissolving process. This is particularly true in the case of highly crystalline samples which can be slow to dissolve. Another useful approach is to use an ultrasonic bath. These provide very powerful agitation and are even more effective when used in combination with a heat source.

2.4.1 Filtration

Of course, there are always samples that refuse to dissolve completely even after ample exposure to heat and prolonged dunking in an ultrasonic bath. Samples that appear in any way cloudy when held up to the light, simply MUST be filtered. Any particulate matter held in suspension will severely compromise field homogeneity and thus line shape (Figure 2.2). Suspended material (of whatever origin) is THE major cause of sub-standard line-shape in NMR spectroscopy.

The whole filtration issue is perhaps a little confusing. Earlier in this section we were stressing the importance of dissolving the *whole* sample and yet here we are, now advocating filtration…? On the face of it, there might seem to be an inherent contradiction in this – and perhaps there is. We can only say that in an ideal world, samples would dissolve seamlessly to give pristine clear solutions without even a microscopic trace of insoluble material in suspension. Samples in the real world are often not quite so obliging! Filtration is very much the lesser of the two evils. If you *know* that you have filtered something from your solution, you are at least aware of the fact that the spectrum is not entirely representative of the sample – but if you *don't*

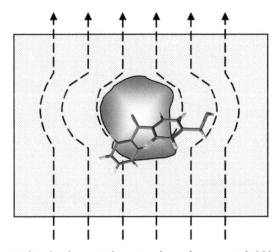

Figure 2.2 Undissolved material causing loss of magnetic field homogeneity.

filter, the resultant spectrum may be so poor as to fail to yield any useful information at all… The choice is as simple as that.

Be warned that very small particle size material, that may even be invisible to the naked eye, is the worst in terms of ruining line shape. The big stuff quite often floats or sinks and therefore doesn't interfere much with the solution within the RF coils.

A convenient method for the filtration of small volumes of liquids is shown in the diagram below (Figure 2.3).

The filter can in a sense be 'customised' as required. A tight plug of cotton wool (rammed hard into the neck of a pipette using a boiling stick) alone is often enough to remove fairly obvious debris from your solution but the addition of a layer of a similarly compacted 'hyflo' on top of the cotton wool makes for a very tight filter which will remove all but the most microscopic of particles. Note that using a pipette bulb to force the liquid through the filter is an excellent idea as it speeds the whole process considerably. Even so, if you are using D_6-DMSO as a solvent, be prepared for a long squeeze as the viscosity of this solvent makes it reluctant to pass

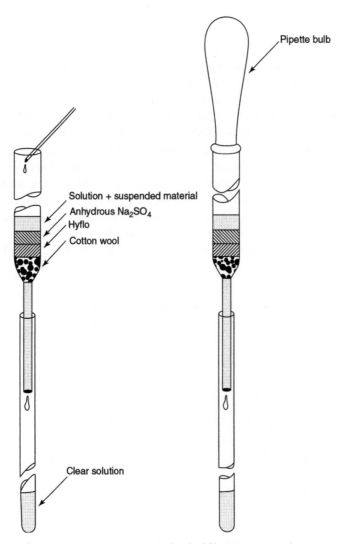

Pipette bulb

Solution + suspended material
Anhydrous Na_2SO_4
Hyflo
Cotton wool

Clear solution

Figure 2.3 A convenient method of filtering NMR solutions.

through a tight filter. If you suspect that your sample is wet (usually, cloudy CDCl₃ solutions with no obvious particulate matter present), you can take this opportunity to dry it at this stage by introducing a layer of anhydrous sodium sulfate to the filter. This will remove most (but not all) of the water present.

A couple of final observations on line shape – just occasionally, we have encountered samples that give very broad lines even after the most stringent filtering. This can be caused by contamination by a tiny amount of paramagnetic material in solution. In one memorable case, a chemist had been stirring a sample around in an acidic solution with a nickel spatula. The tiny quantity of nickel leached from the spatula was sufficient to flatten the entire spectrum. The reason for this is that the ions of any of the transition (d-block) elements provide a VERY efficient relaxation pathway for excited state nuclei, enabling them to relax back to their ground state very quickly. Fast relaxation times give rise to broad lines and vice versa, so to summarise, keep NMR solutions well away from any source of transition metal ions! Should you find yourself in this situation, your only course of action is to run your sample down a suitable ion exchange column.

One other (very rarely encountered) situation is that of the stabilised free radical. It is possible for certain conjugated multi-ring heterocyclic compounds to support and stabilise a delocalised, free electron in their pi-clouds. Such a free electron again provides an extremely efficient relaxation pathway for all nuclei in such a molecule and would give rise to an almost entirely flat spectrum. Such compounds usually give a clue to their nature by being intensely coloured (typically very dark blue). Filtration would do little to improve such a situation but running in the presence of a suitable radical scavenger such as dichloro, dicyano quinone can provide the solution. The scavenger mops up the lone electron and a spectrum can be obtained as normal.

3

Spectrum Acquisition

This was probably the most difficult chapter to put together in the book. For many people who use NMR spectrometers, there will be little (or no) choice about parameters for acquisition – they will probably have been set up by a specialist to offer a good compromise between data quality and amount of instrument time used. This could make this chapter irrelevant (in which case you are welcome to skip it). But if you do have some control over the acquisition and/or processing parameters, then there are some useful hints here. This brings us on to the next challenge for the section – hardware (and software) differences. You may operate a Bruker, Varian/ Agilent, Jeol or even another make of NMR spectrometer and each of these will have their own language to describe key parameters. We will attempt to be 'vendor-neutral' in our discussions and hopefully you will be able to translate to your own instrument's language.

The first thing to note is that there are many, many parameters that need to be set correctly for an NMR experiment to work. Some are fundamental and we don't play with them. Some are specific to a particular pulse sequence and determine how the experiment behaves. It is difficult to deal with all of these here so this chapter will look at some of the parameters that affect nearly all experiments and are often the parameters that you will be able to control in an open access facility. Many of these parameters affect each other and we will try to show where this is the case.

This area is actually quite complex. The descriptions here are not necessarily scientifically complete or rigorous. Hopefully they will help you understand what will happen when you change them (and in which direction to move them).

3.1 Number of Transients

Probably the most basic parameter that you will be able to set is the number of spectra that will be co-added. This is normally called the 'number of transients' or 'number of scans'. As mentioned elsewhere in the book, the more transients, the better the signal-to-noise in your spectrum. Unfortunately, this is not a linear improvement and the signal-to-noise increase is proportional to the square root of the number of transients. As a result, in order to double your signal-to-noise, you need four times the number of transients. This can be shown graphically as in Figure 3.1.

Essential Practical NMR for Organic Chemistry, Second Edition. S.A. Richards and J.C. Hollerton.
© 2023 John Wiley & Sons Ltd. Published 2023 by John Wiley & Sons Ltd.

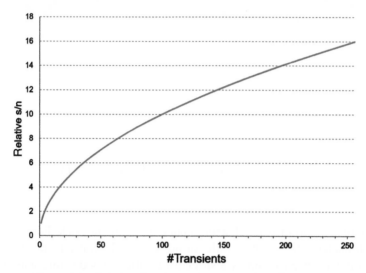

Figure 3.1   Relative signal-to-noise vs. number of transients.

There are several implications of this relationship, the main one being that if you use double the amount of sample, you can acquire the same signal-to-noise spectrum in a quarter of the time. This is particularly apparent if you are acquiring data on insensitive nuclei like ^{13}C where you might be acquiring data for several hours and this can be cut down dramatically if you can spare a little more sample. Don't forget, NMR is a non-destructive technique and you can always get your sample back afterwards (even from DMSO – it just takes a little longer than CDCl$_3$ or MeOD).

Note that often you can't just use any number of transients. Many experiments require a multiple of a base number of transients to work correctly. This is due to the needs of phase-cycling which we won't describe here – once again, check other text books if you want to find out more about this. Generally you will be safe if you choose a multiple of eight as this covers most of the commonly used phase cycles although there are many experiments that can use multiples of two or even one. If in doubt, check the pulse program or ask someone who knows.

3.2   Number of Points

Because the acquisition is digital, you will need to specify how many points you are going to collect the data into. This figure is related to the field that you are operating at – the higher the field, the more points that you are going to need. This parameter relates to the spectral width observed and the acquisition time through sampling theory. The Fourier transform algorithm demands that the number of points is a power of 2 so we tend to use the computer term of 'k' to describe the number of points (where 1 k = 1024 points). If we acquire 20 ppm at 400 MHz, this has a spectral width of 8000 Hz. If we then want to have a digital resolution of 0.5 Hz we would need 16 k points to achieve this. Because we are acquiring both real and imaginary data, we would need to double this so we would need 32 k points to achieve this resolution. We can improve the appearance a little by using 'zero filling' and this is described later.

3.3 Spectral Width

Sampling theory states that you must sample a waveform at least twice per cycle otherwise you will observe a lower-frequency signal than the true signal (you often see this effect in old cowboy films where wagon wheels speed up and then appear to stop and move backwards). In NMR we set the sampling rate by choosing the spectral width of the spectrum. If we choose too narrow a spectral width, then signals outside that range will 'fold back' into the spectrum (normally with strange phase). Older spectrometers use electronic filters to try to avoid this but even the best ones don't cut off frequencies exactly; they attenuate signals close to the filter edge. More modern spectrometers use oversampling and digital filters which treat the spectrum computationally to produce a very sharp filter. This means that folding ('aliasing') is seldom seen in 1-D spectra. This is not the case in 2-D spectra though, as the indirect dimension cannot benefit from these filters. In this case, setting too narrow a spectral width in the second dimension will result in folded peaks in the 2-D spectrum.

3.4 Acquisition Time

This parameter is not normally set directly but is a function of the values that you set for spectral width and number of points. The narrower the spectral width, the longer will be the acquisition time and the greater the number of points, the longer the acquisition time.

3.5 Pulse Width/Pulse Angle

When we excite the nuclei of interest, we use a very short pulse of radiofrequency. Because the pulse is very short, we generate a spread of frequencies centred about the nominal frequency of the radiation. The longer this pulse, the more power is put into the system and the further that it tips the magnetisation from the Z-axis. We call this the 'flip angle'. A 90° flip angle gives rise to the maximum signal (you can picture it as the projection on the X-Y plane, where Z is the direction of the magnetic field). This is shown diagrammatically in Figure 3.2.

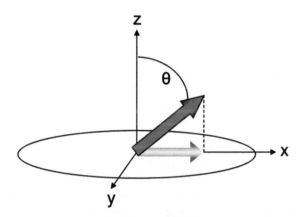

Figure 3.2 'Flip angle'.

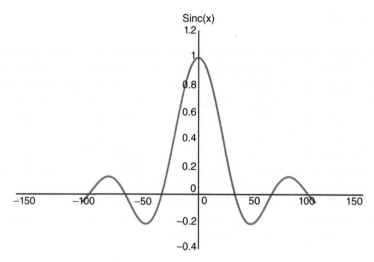

Figure 3.3   The 'sinc' function.

The other consequence of the pulse width is the spread of frequencies generated. The shorter the pulse, the wider will be the spread of frequencies. Because we often want to excite a wide range of frequencies, we need very short pulses (normally in the order of a few microseconds). This gives rise to a so-called sinc function (Figure 3.3).

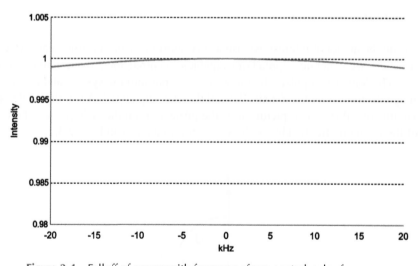

Figure 3.4   Falloff of power with frequency from central pulse frequency.

At first sight, this may appear to be a lousy function to excite evenly all the frequencies in a spectrum but because we use such a short pulse, we only use the bit of the function around $x = 0$. The first zero-crossing point is at $1/(2 \times$ pulse width) – this would be at about 150 kHz for a 3 μs pulse. For a 400 MHz spectrometer, we need to cover a bandwidth of about 8 kHz for a proton spectrum. As Figure 3.4 shows, there is minimal power fall off for such a small pulse.

Of course, it is quite easy to solve the bandwidth needs of proton spectra – they only have a spread over about 20 ppm (8 kHz at 400 MHz). Things get a bit more difficult with nuclei such as ^{13}C where we need to cover up to 250 ppm (25 kHz) spread of signals and we do notice some falloff of signal intensity at the edge of the spectrum. This is not normally a problem as we seldom quantify by ^{13}C NMR. However, it can be a problem for some pulse sequences that require all nuclei to experience 90° or 180° pulses. This is particularly true at higher fields but we now have access to different ways of generating these transitions using so-called adiabatic pulses.

One last comment about pulse widths; it is important that we know what the 90° pulse width is for the nuclei that we observe as accurate pulse widths are required for many pulse sequences (as mentioned previously). Failure to set these correctly may give rise to poor signal-to-noise or even generate artefacts in the spectrum. When instruments are serviced, these pulse widths are measured and entered into a table to ensure that the experiments continue to work in the future.

3.6 Relaxation Delay

This is the amount of time included in a pulse sequence to allow all the spins to lose their energy. Failure to let this happen will cause some signals to integrate less than they should (or may cause artefacts in some experiments). The amount of time that you leave depends on the amount that you have tipped the spins with your excitation pulse (see 'pulse width'). If you have made a 90° pulse then you will have to wait for about 30–50 seconds between pulses to allow the spins to re-equilibrate. The exact length of time is specific to the environment of the nuclei that you are observing. Generally, singlets are the slowest signals to relax and will tend to under-integrate if you have too short a relaxation delay. The spins have the total time from when they were excited until their next excitation to relax. This means that the value that you set for the relaxation delay also depends on the acquisition time.

For most 1-D proton experiments we tend to use a pulse angle of about 30° and an acquisition time of about 3 seconds – so a relaxation delay of about 2 seconds is normally fine for most proton work. If you need super-accurate integrals you can play safe and give a relaxation time of 10 seconds and this should cover most eventualities. So why not just set a relaxation delay of a minute? This would obviously cover every eventuality. The problem is that this delay is inserted into every pulse cycle so your experiment would take a long time to complete. It ends up that you have a compromise of how much you tip the spins, how long you acquire for and how long you wait for. For example, if you get maximum signal by using a 90° pulse you may have to wait such a long time for the spins to relax that you don't achieve the throughput that you were after. It turns out that the optimum flip angle (in terms of rate of data collection) is about 30° and this is what we use for most 1-D proton spectra.

3.7 Number of Increments

For 2-D experiments, not only will you need to set the number of points for your direct detection dimension, you will also need to set the number of experiments in the second dimension as this will determine what resolution you have in that dimension. There is no simple answer to help here – it depends on the experiment that you are performing, what information you need and what frequency you are operating at. For a COSY experiment, we probably need quite a few increments because we are often interested in signals that can be quite overlapped. In this case, 256 increments would be quite reasonable at 400 MHz. If you were operating at 800 MHz then you would need double this (512 increments) to get the same resolution. There are some mathematical tricks that we can perform with the data to improve this situation and these are described in the next chapter. By the way, you have to be a bit careful with the name for the second dimension – Bruker call it 'f1' and Varian call it 'f2'. In this book we will stick with the term 'the indirect dimension'.

3.8   Non-Uniform Sampling (NUS)

It turns out that it is possible to derive the necessary information in a 2-D (or greater dimensionality) spectrum without having to sample the second dimension sequentially with a defined increment. There are different approaches to sampling the second dimension, each with their own advantages and disadvantages. You can select randomly, or exponentially or using fancy schemes (e.g. sine-weighted Poisson-gap sampling). One thing that is in common with all of these schemes is that you cannot perform a simple Fourier transform on the data in the second dimension because the Fourier transform requires data to be sampled uniformly. The NUS approach requires a specialist transform which understands the sampling method (see 2-D section in Chapter 8).

3.9   Shimming

When we are looking at NMR data, we need to be able to resolve 0.5 Hz (or better) in a few hundred MHz. This means our field must be homogenous to better than 1 part per billion! Magnets are created to exacting standards and produce highly linear fields in the sample area. This is achieved through very precise engineering and the addition of coils which can tweak the field to make it even more precise (by passing a current through them). These coils are called 'shim coils' and they come in two different flavours: cryoshims and RT (room temperature) shims. The cryoshims are at liquid helium temperatures and are set up when the magnet is energised. The cryoshims are capable of getting the field homogeneity to better than 5 parts per million and once they are set up they are not normally altered. To get the field to the desired homogeneity we use the RT shims and these are adjusted by passing different amounts of current through them. Changes in the environment due to the sample or other external factors may cause this field to be distorted. To get the field to the ultimate homogeneity, the RT shims are adjusted so that they contribute field to add or take-away from the main field. There are a large number of these shim coils (up to 40 on some magnets) and they each have a particular influence on the magnetic field. They are named after the mathematical function of the field that they supply. The basic ones are somewhat obviously called 'X', 'Y' and 'Z'. That is, they have a linear effect on the field in the X, Y and Z directions. The more complex shaped ones have esoteric names such as X2Y2Z4 ($X^2Y^2Z^4$).

So, given this frightening range of coils, how do you go about shimming a system? The answer is: 'with lots of experience'. To be able to shim a system from scratch is a highly skilled job and requires huge patience. Fortunately, you may never have to do it. Once a system is set up, the shim values (how much current is passing through each shim coil) for most of the shims remain relatively static. We normally only have to tweak the 'low-order' Z shims in daily use. This means 'Z' (nearly always), 'Z2' (nearly always), 'Z3' (quite often) and 'Z4' (sometimes). The rest, we can normally ignore. Modern spectrometers will go even further to help you and will shim your sample automatically. This normally uses a 'simplex' approach and takes about a minute or two. In addition to this, there is the more recent development of 'gradient shimming'. Unlike the simplex method (which gives the coil a tweak and looks at the result and then decides what to do next), gradient shimming acquires a map of the field and then works out which functions will make it more homogeneous. It then sets the values in the coils and doesn't have to go through the iterative process of the simplex method. The simplex method takes longer for each shim that you optimise whereas the gradient approach will take the same length of time to do all of the 'Z' coils.

Shimming is not yet a 'thing of the past' but it is certainly less of a badge of honour for budding spectroscopists.

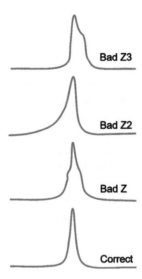

Figure 3.5   Correct line shape and some typical distortions caused by poor shimming.

Sometimes the (automatic) shimming process goes wrong and the instrument is unable to generate the field homogeneity that is needed. You will need to spot this otherwise you may make the wrong judgement about your compound. So how can you tell? Well, the key is to understand what physically happens if the field is not homogeneous. Your sample should experience the same strength field wherever it is in your sample tube. If it doesn't, then molecules in different parts of the tube will resonate at slightly different frequencies. This will give rise to line broadening and, depending on the shim which is out, may give rise to distinctive line shapes. Some of the common distortions are shown in Figure 3.5.

As mentioned earlier, poor lineshape may be due to a number of different factors and these are covered in the sample preparation chapter. It is important to know whether the poor lineshape is due to sample or spectrometer – after all, you don't want to spend time playing with your sample when the spectrometer was the problem all along and conversely, you don't want to spend time fruitlessly shimming the spectrometer when the problem lies with your sample. Dynamic sample effects can be identified because the sample signals will be broad but the solvent (and impurity signals should there be any) will be sharp. It is more difficult to distinguish sample preparation effects from shimming effects as they both affect all signals in the spectrum. The smoking gun for an instrumental problem is if the samples before or after yours also look bad. If they are fine, it's probably your sample. If they are bad then it's probably the instrument (unless you prepared them, in which case it could be your technique). In the case of the 'Z' shim, you may end up with multiple peaks for each of your real peaks. If you don't realise that this is a shimming problem then you might assume that your sample is impure when it is not. Note, however, that bad shimming is unusual so don't use it as an excuse to pretend that your compound is pure when it is really a mess. You can always check – look at the solvent peaks in the spectrum. If they are split too, then it is shimming – if they aren't, it's your sample! Note that no amount of shimming, manual or automatic, can compensate for undissolved material in solution, or incorrect sample depth.

In conclusion, shimming is best left to the experts (or the instrument) but it is important to be able to spot shimming problems so that you don't misjudge your sample.

3.10   Tuning and Matching

The NMR probe is a tuned radiofrequency circuit. When we insert a sample in the coils of the probe, we affect the circuit and can change its resonant frequency. If the circuit becomes de-tuned, it becomes less efficient at transmitting the radiofrequency to the sample. This often results in pulses that do not tip the magnetisation as much as we were hoping to achieve. As mentioned earlier, this can have a detrimental effect on complex pulse sequences and create artefacts in the spectrum (or decrease the signal-to-noise for simple pulse sequences). Tuning and matching allow us to tweak the circuit to compensate for the sample load on the coils. In older systems (many of which are still in use), tuning and matching is carried out on the probe using tuning knobs. In more modern systems this is done under automation by the instrument. Differences in probe tuning can be seen when running different solvents after each other (e.g. $CDCl_3$ followed by DMSO) or if you have 'lossy' samples which are highly conductive (e.g. salt solutions).

3.11   Frequency Lock

Because the magnetic field of an NMR spectrometer can drift slowly over time, it is necessary to 'lock' the spectrometer frequency to something that drifts at the same rate. This is achieved by monitoring the deuterium signal in your solvent. As the magnet field drifts, so does the deuterium signal and this moves the spectrometer frequency at the same time. Normally you don't need to think about this but it becomes important when you are using a mixed solvent as the instrument may lock onto the wrong solvent signal. If this is the case, your chemical shifts will be incorrect. You can check whether this has happened by looking at your residual solvent signals (or TMS if you have any in your sample).

Obviously, if you are running in a non-deuterated solvent you will not be able to lock your sample. In this case there are a few options:

3.11.1   Run Unlocked

If your experiment is short, you don't need to worry about field drift. Modern magnets are quite stable and can be used for at least a few minutes without drifting too far. The disadvantage with this approach is that shimming is normally performed on the deuterium signal and you will need to shim your sample differently if there is no deuterium in the sample.

3.11.2   Internal Lock

You can keep the spectrometer happy by adding a deuterium source to the sample. On the other hand, you probably don't want to do this otherwise you would have selected a deuterated solvent in the first place! Nonetheless it is still an option in some cases. Note that if you only add a small amount of the deuterium source, you may struggle to achieve lock because the signal is too weak.

3.11.3   External Lock

If you don't want to contaminate your sample, you can use a small tube, filled with a deuterated solvent, inside your main tube, as mentioned earlier. This is particularly useful if you are running a neat liquid or if the deuterated solvent is immiscible with your sample. This is probably the most common approach to this problem.

Finally, it has been noted that some people think that they need TMS in their samples to enable them to lock. This is not the case. On modern spectrometers, TMS is used for referencing only. There was a time when it was used for locking CW instruments (in an early form of spectrum averaging) but it is not used in that way for FT instruments now.

3.12   To Spin or Not to Spin?

In the early days of NMR, spinning the sample was seen as essential. The reason for spinning is to average out inhomogeneity in the magnetic field which can be caused by the sample or poor shimming. By rotating the sample tube, molecules will experience an average field. This can improve the resolution of the signals which is obviously a good thing. With modern NMR systems, however, this is seldom necessary. Magnetic field homogeneity has improved considerably over the years due to better magnet design, shim system design and shimming software. Spinning is not without its problems, particularly in very sensitive probes, and can introduce its own artefacts such as Q-modulation sidebands in 1-D spectra (antiphase peaks either side of the main peak) and other artefacts in 2-D spectra.

The advice for most modern spectrometers is not to spin. A little time spent in decent sample preparation should make this unnecessary. From experience in the real world, we have found that sample preparation is not always of the highest standard and spinning may help to correct this to some extent. In the end, for a workhorse 400 MHz system with an ordinary probe, it is a pragmatic decision based on your individual needs. If you are lucky enough to have a high performance probe then it is best not to spin.

4

Processing

4.1 Introduction

Acquiring your data is just the first step in producing a useful spectrum. Fortunately, systems are normally set up so that they perform the processing steps automatically. Most of the time they do an excellent job and your data is fine. Sometimes you may have special requirements and other processing will be required. This chapter looks at some of the things that can be altered to improve the appearance of the data for you. Note that most of the examples are for 1-D proton spectra but all of the sections are valid for certain types of 2-D experiment.

4.2 Zero-Filling and Linear Prediction

Because we are always in a hurry (so many samples, so little spectrometer time) we always try to acquire that little bit faster than we should. This is particularly true with 2-D acquisitions which can be very time-consuming. As discussed previously, we try to minimise the number of increments to save time. This gives rise to highly truncated data sets and poor resolution. This can be made to look a little prettier by adding a load of zeros to the experiment before Fourier transforming it. We call this (somewhat obviously) 'zero-filling'. Note that this doesn't add any information but it does make the result look nicer.

Linear prediction works in a different way by predicting what the missing (future) values would be, based on the existing (past) values. This approach is more powerful than mere zero-filling but it also brings with it some risks (artefacts). You can't linear predict infinitely and so we tend to advise that one degree of linear prediction is about all the data can reliably take without going into the realms of fantasy. If we take the example of our COSY spectrum, we would probably linear predict out once (to double its size to 512 points) and then zero fill once or twice to take the final size to 1024 or 2048 points (in the indirect dimension). It is also possible to 'backward linear predict'. This allows us to reconstruct the first part of the FID which we can't collect because we have to wait a finite time for the powerful excitation pulse signal to die away. This effect is known as 'ring down' and causes baseline distortion. Backward linear prediction allows us to throw these points away and replace them with what might have been there.

Essential Practical NMR for Organic Chemistry, Second Edition. S.A. Richards and J.C. Hollerton.
© 2023 John Wiley & Sons Ltd. Published 2023 by John Wiley & Sons Ltd.

4.3 Apodization

Sometimes the FID doesn't behave in the way that we would like it to. If we have a truncated FID, Fourier transformation (see Section 4.4) will give rise to some artefacts in the spectrum. This is because the truncation will appear to have some square wave character to it and the Fourier transform of this gives rise to a sinc function (as described previously). This exhibits itself as nasty oscillations around the peaks. We can tweak the data to make these go away by multiplying the FID with an exponential function (Figure 4.1).

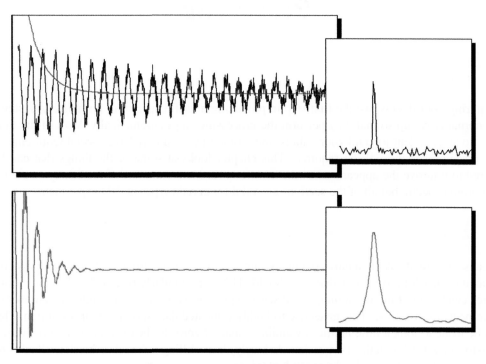

Figure 4.1   Exponential multiplication.

This has the effect of smoothing the FID away to zero, thus yielding lovely peaks. We call this 'exponential multiplication' for obvious reasons!

It is also possible to play other mathematical tricks with the FID. For example, we may want to make our signals appear sharper so we can see small couplings. In this case, we want our FID to continue for longer (an infinite FID has infinitely thin lines when Fourier transformed). To do this we use 'Gaussian multiplication'. This works exactly the same way as exponential multiplication but uses a different mathematical function (Figure 4.2).

It should be noted that Gaussian multiplication can severely distort peaks and also reduce signal-to-noise of the spectrum so it is not a good idea to do this if you have a very weak spectrum to start with. Spectrum 4.1 shows a real case where Gaussian multiplication has been able to resolve a doublet of doublets.

There are many other apodization functions which are used for specific types of NMR data. In fact you can make up your own if you want to but for most data sets, the 'canned' ones that are shipped with the instrument are more than adequate.

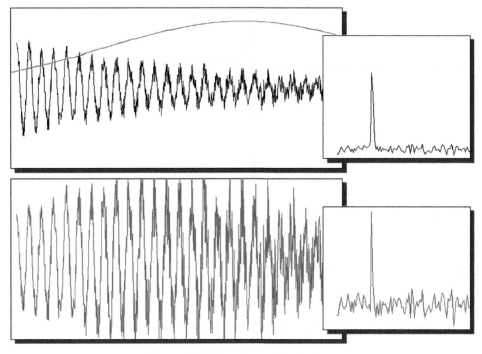

Figure 4.2 Gaussian multiplication.

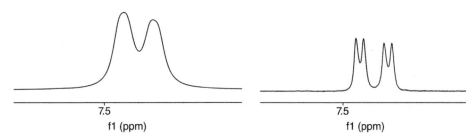

Spectrum 4.1 Gaussian multiplication in action.

4.4 Fourier Transformation

As mentioned earlier, we acquire data in the 'time domain' but to make sense of it, we need to view it in the 'frequency domain'. This is where the Fourier transformation comes in. There is not too much to do here – there are no parameters to change. It is a necessary step but the automatic routines will perform this for you with no input.

4.5 Phase Correction

For several technical reasons, it is not possible to acquire NMR data with perfect phase. One reason is the inability to detect XY magnetisation correctly; another is the fact that we are unable to collect the data as soon as the spins are excited. These limitations mean that we have

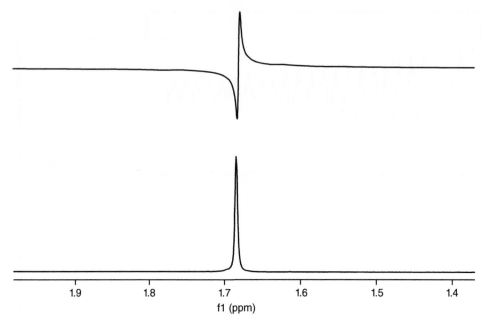

Spectrum 4.2   An absorption signal (below) and dispersion signal (above).

to phase correct our spectrum so that we end up with a pure absorption spectrum. What we *don't* want is a dispersion signal – see Spectrum 4.2.

The XY problem gives rise to a constant phase error across the spectrum, the delay problem gives a linear phase error. To correct for this, we have two phase adjustment parameters at our disposal: zero and first order.

Modern NMR software comes with very good automatic phase routines so most of the time you should end up with a beautifully phased spectrum. Sometimes, however, the software doesn't quite perform and you may need to tweak the phase manually. It can take a bit of familiarity to get this right but it is just a matter of practice. If you remember that the zero-order adjustment works constantly across the spectrum and that the first-order doesn't, it is quite easy to see what is going on. Normally the software gives you an option of setting the 'pivot point' of the first-order adjustment (i.e. the frequency in the spectrum where there is no effect from the first-order adjustment). This pivot point is normally set to the largest peak.

Spectrum 4.3 shows how the phase can be improved with a manual tweak. Note that in a poorly phased spectrum, the integrals will be distorted such that they are pretty much unusable.

So far, we have talked about phasing 1-D spectra but this is also valid for some 2-D experiments. Phase-sensitive 2-D experiments also require phasing in one or both dimensions. Similar approaches are used as described here. Note that this is not necessarily the case for all 2-D experiments as some of them are collected in 'magnitude mode' where we look at only the intensity of the signals, not their sign.

One last CAUTIONARY NOTE: the first-order phase can be increased beyond $\pm 360^0$ – but shouldn't be! If this happens, you will end up with a distorted, 'wavy' baseline. A sine wave is in effect superimposed on the spectrum, so if you see a wavy baseline, check that you

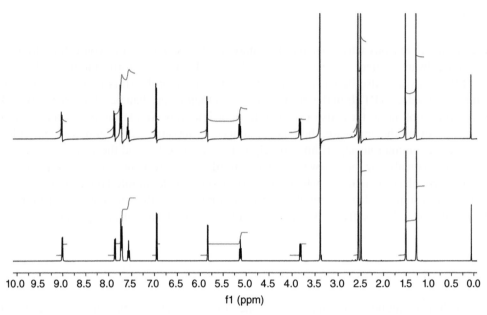

Spectrum 4.3   A well-phased spectrum with reliable integrals (below) and a badly phased spectrum with unusable integrals (above).

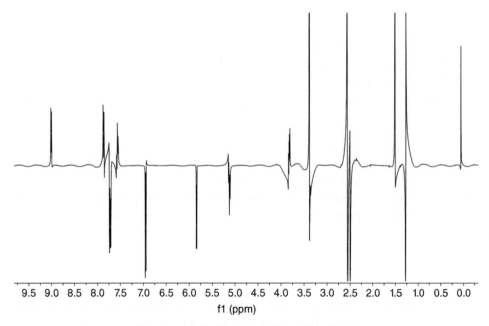

Spectrum 4.4   Too much first-order phase!

haven't wrapped the phase too far. Spectrum 4.4 shows what happens when you go a bit mad with first-order phase! If you end up in this position, do not attempt any kind of baseline correction as this will add to your problems. Just set both your phase parameters back to zero and start again…

4.6   Baseline Correction

There are many reasons why your baseline may not be as flat as you would like. Many of them are hardware-related; some are brought about by having a distortion in the early points in the FID. They can also be caused by background in the probe (this is often the case for fluorine spectra due to PTFE in the probe). Whatever their cause, bad baselines not only make the spectrum look poor, they also give rise to poor integrals. While there has been a lot of work at improving the hardware, there is still a need to massage the baseline to make it look good. There are numerous algorithms to help with your baseline and these will generally be applied automatically by the software that was used to acquire your data. These poor baselines are particularly noticeable when you have a very weak sample (for example, carbon spectra). It is also possible to manually correct your baseline if the automatic algorithms fail. In this case, you tell the software where the baseline should be and it then performs a spline-fit to level it.

4.7   Integration

As you will no doubt be aware, integrals are one of the key parameters in the interpretation of proton spectra and are pivotal in quantification. They measure the area under a peak and this is directly proportional to the number of protons (in the case of proton NMR) in that environment. Most software will automatically try to identify the peaks in your spectrum and integrate them for you. If you need to do it yourself, then it is a fairly trivial matter of defining the start and end point of the integrals of interest. The only complication is that you may need to tweak the slope and bias of the integral. This should be unnecessary if you have got the phase and baseline of your spectrum correct. If you find that you need to adjust slope and bias, we suggest that you go back and try to sort out baseline and phase a bit better.

Integrals may appear low on signals that have a long relaxation time (see Chapter 3). If this is the case, then you should acquire your data with a longer relaxation delay. This is likely to be most noticeable on singlets and isolated protons as these tend to have quite long relaxation times. If you have poor signal-to-noise, this will also affect the accuracy of your integrals.

Lastly, be aware that a signal spreads quite a long way from the centre of the peak. This is particularly true in the case of broad peaks like exchangeable signals. To get a good integral for an exchangeable you need to set your integration width very wide. Most automatic integration routines do this badly, so expect the automatic integrals to integrate low for these types of peaks.

4.8   Referencing

As mentioned in Chapter 3, we standardise our reporting of chemical shifts with reference to TMS or the residual solvent peak. Your spectrometer software should do this for you automatically. If it gets it wrong (which is possible if you have a mixed solvent or a spurious peak near TMS), then you can set it manually using your software.

4.9   Peak Picking

If you want accurate chemical shifts or splittings, peak picking can help. However, it is worth issuing a health warning here! The accuracy of your chemical shifts and your splittings is limited by the digital resolution of your spectrum. This means that while the computer is happy to spout figures to four decimal places, in reality you may not be able to measure to better than ±0.5 Hz. Always check your digital resolution before trying to quote things too accurately. Don't forget, your chemical shifts will be influenced by concentration, temperature and pH so it is probably pointless quoting chemical shifts to a greater accuracy than 0.05 ppm except in special circumstances. Also, be warned that measured splitting is influenced by line width so very broad peaks (or very close peaks) may show a smaller value than the real value.

5

Interpreting Your Spectrum

We should perhaps make a few important points before going any further – the title of this chapter is highly ambitious! We certainly cannot promise to turn the inexperienced reader into expert interpreters in the time it takes to read this section. Experience is essential and to become really proficient in this area, you need to critically examine literally *thousands* of spectra. However, be that as it may, by establishing some sound principles and cultivating a critical approach to the spectra you encounter, this book should prove useful in helping along the way.

It might be worth considering at this stage, what we really mean by the term 'spectral interpretation'. What do we consider to be acceptable criteria for the interrogation of spectral data? Is a cursory glance sufficient if you are also holding a mass spectrum in your other hand that shows a parent ion of the correct mass for your desired compound? Or should you throw every known NMR technique at all your compounds irrespective of how trivial the chemical change being attempted? These questions should be pondered in the light of the fact that an NMR spectrum should never be regarded in itself as an *absolute* proof of structure. If this is what is required, then you had better practice your crystal-growing skills because you will be needing the services of an X-ray crystallography department. That having been said, NMR data can certainly provide the next best thing – in the right hands…

Our initial observations are aimed at improving your understanding of 1-D proton spectra, though many of the principles we will try to establish will be equally applicable to other nuclei too. We will discuss issues specific to ^{13}C interpretation, later. Please note that there are many other sources of help in interpreting your spectrum, not least on the internet. Please be very careful with these as we have observed a large number of them which are simply wrong and obviously presented by people who have, at best, an elementary grasp of NMR (at worst, a totally flawed understanding!).

As we mentioned in the Introduction, it is ironic that one of the major problems encountered when dealing with NMR spectra, is the sheer quantity of information that you are presented with. Unless you are practised in the art of interpretation, you may find yourself swamped by it. Clearly, a methodical and universally applicable approach would be advantageous. There is not necessarily a 'right' or a 'wrong' way to approach a spectrum, but some ways are probably better than others! These are our 'Top Ten' recommendations, for what they are worth.

Essential Practical NMR for Organic Chemistry, Second Edition. S.A. Richards and J.C. Hollerton.
© 2023 John Wiley & Sons Ltd. Published 2023 by John Wiley & Sons Ltd.

1. Take a moment to survey the spectrum and ask yourself if it is fit for purpose? Of course, if you have run it yourself, then it should be fine but this may not always be so with walk-up systems. Is the line shape and resolution up to standard? Has the spectrum been phased correctly? Is the vertical scale well-adjusted so that you can see the tops of all the peaks (except perhaps, obvious singlets)? Are the integrals well displayed? If the horse is dead, don't flog it – get a new one. (Note: A good walk-up system will run day and night, producing quality results for the vast majority of samples. However, the occasional spectrum may 'come off the rails' for no obvious reason but remember that there are dozens of processes that must run correctly in the background before a high-quality spectrum drops into the collection tray and a slight hiccup in any of them can spoil the end result. Some of these problems (vertical scaling, phasing and integration) can be rectified by reprocessing the acquired data and some cannot as the 'raw' data itself may be unsuitable (poor signal/noise and sub-standard shimming).)

2. If the spectrum is satisfactory, you can get to work on it. Can you identify any obvious impurities or solvents that might be present? Crossing them off at this stage is a valuable exercise in data reduction and clears the way ahead so that you can concentrate on the important peaks.

3. Does your proposed structure exhibit any special features likely to have a significant effect on your spectrum? (E.g. chiral centres, sites of potential restricted rotation, abnormal stereochemistry etc.)

4. Can you identify a signal that gives a clear integration for a known number of protons?

5. Now work from left to right, assigning each signal, or groups of signals that you observe, to protons in your proposed structure. (If there is logic in starting at the left of the spectrum, it is that most molecules have some aromatic or heterocyclic core, to which, various alkyl functions are attached. If there is a problem with the core, then you will at least discover it promptly and be able to relate it to the alkyl components of the molecule.)

6. Interrogate each and every signal in your spectrum to check that they conform to the expected values for the **three** crucial NMR parameters – **Chemical Shift, Coupling pattern** and **Integration.** And this, in a very real sense, must form the basis of our working definition of 'interpretation'. Two out of three might be good enough for Mr. Meatloaf, but it just doesn't cut the mustard for any spectroscopists worth their salt! Obviously, in order for you to match your observed values for chemical shifts and couplings, to expected values, you will need a great deal of data at your disposal and this will be provided in the following chapters…

7. If you note an obvious mismatch between observed and predicted data, might you have overlooked something in (3) above? Interpretation is essentially an iterative process. Try to maintain a degree of flexibility in your approach – without being *too* flexible! Achieving this balance takes practice! If there is still no way of reconciling observations with predictions, you must accept the strong probability that your proposed structure is incorrect.

8. If so, propose an alternative structure and start the whole process again. Alternatively, could your sample be a mixture? If so, might your sample benefit from a chromatographic investigation at this stage or is it possible to qualify and quantify the components directly?

9. If you have any reasonable cause for doubt (e.g. because some key signals in your spectrum are obscured etc.), would the acquisition of more NMR data be helpful? If so, consider exactly what you wish to achieve and select the appropriate technique and gather the data.

10. Re-evaluate all data again and again until you are as happy as possible with all aspects of your spectrum. Guard against complacency! Is it watertight? Check on this by asking yourself if you would be happy to stand up in a court of law and defend your efforts.

Adherence to all these points might seem to make the whole process of interpretation incredibly convoluted and unappealing, but in reality, it should eventually become 'second nature'. Developing the theme further, you will hopefully soon get used to mentally synthesising the spectra of compounds you look at and matching these against what you see before you. The degree of deviation between the two will be critical and we will explore this a little later. Should you be lucky enough to have a job which involves you looking at literally, *thousands* of spectra, then mental short-cuts will evolve and you will find yourself taking-in and digesting certain patterns almost subliminally in the same way that you read the words on this page. This is perfectly valid but **always** be on your guard against complacency! Take pride in your craft. *Make* yourself re-visit that splitting pattern. There is nothing worse than falling into an obvious trap!

Of course, we do not mean to imply any kind of moral imperative here. For example, if you were checking out some dubious looking starting material that supposedly contained a given functionality (e.g. Ar–CH_2–Br with no ortho-substituent on the aryl ring and no other obvious steric clash between this aryl system and any other), then it would be quite acceptable – and even desirable – to exploit some obvious 'handle' in the form of a 'cut-down interpretation'. In this case, for example, if there is no two-proton singlet (or AB system if the molecule is chiral) at around 4.6–4.3 ppm, then you can reasonably conclude that the stuff is not what you wanted. Job done. Move on.

We discussed earlier, the concept of mentally synthesising a spectrum and trying to match it with real, observed data. This might sound like a task that a computer would be very good at – but this is certainly not the case in the field of proton NMR! We will discuss this in detail in Chapter 15 dealing with software issues. Our synthesised spectrum will need to be based on hard data. For example, in the case above, just how much leeway can you allow yourself in predicting the chemical shift of the Ar–CH_2–Br protons? ±0.1 ppm? ±0.2 ppm? More? Instinctively, of course, the closer our observed signals are to our predicted ones, the better we will feel about them and vice versa. We have tried to portray this in Figure 5.1.

While this might seem to state the obvious, it begs a number of very important questions. Ultimately, and perhaps after the expenditure of a great deal of mental effort, the bottom-line of an interpretation boils down to a simple question: Does this spectrum support this putative structure or doesn't it? Yes or no? The question is clearly black or white but the answer has plenty of scope for shades of grey! The yes/no question will have to be answered on the back of a great many other questions regarding the 'goodness of fit' of all the signals in the spectrum.

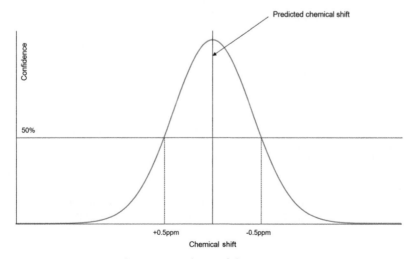

Figure 5.1   'The confidence curve'.

Let's return to the simple case of Ar–CH_2–Br above, and how far the observed shift can be allowed to deviate from our predicted position. At what stage must we reject the structure? After all, how can we be sure that it isn't Ar–CH_2–Cl instead? Or what if the aryl ring is further substituted? What would you expect to see if there was a –NO_2 ortho- to the –CH_2–Br? Or what about a para- –NR_2?

Hopefully, we make our point. The confidence curve is NOT fashioned out of granite – it has to be applied with understanding and circumspection. It will always have the same basic shape but we have to be prepared to take a view on how wide it should be in every individual situation! Matters become a great deal more complex when we come up against structures that are sterically crowded i.e. structures where bond constraints force various moieties into close proximity with one another.

(Note also that the concept of the 'confidence curve' is equally applicable when considering coupling data. That is, what size coupling should I be looking for in this system or that? Is it too big? Too small?)

Unfortunately, it is impossible to cover all the potential pitfalls that wait for the unwary. Many more will come to light in the following chapters but for now we will concentrate on supplying you with useful proton NMR chemical shift data… We have done this by collating various types of protons into convenient 'groups' but firstly, let's clear the wood from the trees and deal with commonly encountered solvents and impurities in the regularly used NMR solvents.

5.1   Common Solvents and Impurities

As we pointed out earlier, it is a good idea if you can eliminate peaks from common solvents and impurities before getting into the real interpretation (note how chemical shifts can vary in different solvents – another factor which helps define the breadth of the confidence curve…). Table 5.1 can be very helpful in this regard.

Table 5.1   The proton chemical shifts of common solvents and impurities.

Impurities	CDCl$_3$	DMSO	D$_2$O	MeOD
Acetic acid	2.13	1.95	2.16	1.99
Acetone	2.17	2.12	2.22	2.15
Acetonitrile	1.98	2.09	2.05	2.03
Benzene	7.37	7.40	7.44	7.33
Bromoform	6.85	7.75	Insoluble	7.42
n-Butanol	3.67(t,6)	3.41(t,6)	3.60(t,6)	3.54(t,6)
	0.94(t,7)	0.89(t,7)	0.89(t,7)	0.93(t,7)
t-Butyl alcohol	1.28	1.14	1.23	-
Chloroacetic acid	4.14	4.28	4.25	-
Chloroform	7.27	8.35	Insoluble	7.88
Cyclohexane	1.43	1.42	Insoluble	1.45
1,2-Dibromoethane	3.63	3.84	3.79	3.72
Dichloroacetic acid	5.98	6.68	6.21	-
1,2-Dichloroethane	3.73	3.93	3.92	3.78
Dichloromethane	5.30	5.79	Insoluble	5.48
Diethyl ether	3.48(q,7)	3.42(q,7)	3.56(q,7)	3.48(q,7)
	1.20(t,7)	1.13(t,7)	1.17(t,7)	1.17(t,7)

Table 5.1   (Continued)

Impurities	CDCl$_3$	DMSO	D$_2$O	MeOD
Diisopropyl ether	1.12(d,6)	1.04(d,6)	1.12(d,6)	-
Dimethylacetamide	3.01	2.99	3.05	3.05
	2.94	2.82	2.89	2.91
	2.08	1.99	2.08	2.07
Dimethylformamide	8.01	7.98	7.91	7.98
	2.95	2.92	3.00	2.99
	2.88	2.76	2.86	2.85
Dimethyl sulfoxide	2.62	2.52	2.70	2.65
Dioxan	3.70	3.61	3.75	3.65
Ethanediol	3.76	3.42	3.66	-
Ethanol	3.72(q,7)	3.49(q,7)	3.64(q,7)	3.60(q,7)
	1.24(t,7)	1.09(t,7)	1.16(t,7)	1.17(t,7)
Ethyl acetate	4.12(q,7)	4.08(q,7)	4.14(q,7)	4.09(q,7)
	2.04	2.02	2.08	2.01
	1.25(t,7)	1.21(t,7)	1.23(t,7)	1.23(t,7)
Ethyl formate	8.04	8.23	8.16	-
	4.22(q,7)	4.17(q,7)	4.28(q,7)	-
	1.29(t,7)	1.24(t,7)	1.29(t,7)	-
Formic acid	8.02	8.18	8.22	-
Isobutyl methyl	2.12	2.08	2.19	2.11
ketone	0.92(d,6)	0.88(d,6)	0.88(d,6)	0.91(d,6)
Isopropyl acetate	2.02	2.00	Insoluble	1.99
	1.22(d,6)	1.21(d,6)		1.22(d,6)
Isopropyl alcohol	1.2(d,6)	1.06(d,6)	1.18(d,6)	1.14(d,6)
	4.03(m)			3.92(m)
Methanol	3.48	3.20	3.35	3.35
Methyl acetate	3.67	3.61	3.68	-
	2.05	1.92	2.09	
Methyl iodide	2.16	2.21	Insoluble	2.15
Morpholine	3.69(m)	3.52(m)	3.70(m)	3.64(m)
	2.85(m)	2.68(m)	2.79(m)	2.79(m)
Nitromethane	4.32	4.44	4.41	-
Petroleum spirit	1.28	1.28	Insoluble	1.30
(60°–80°)	0.90	0.89		0.88
Potassium Acetate	insoluble	1.60	1.91	-
Propanol	3.60(t,7)	1.45(m)	3.61(t,7)	3.49(t,7)
	1.60(m)	0.87(t,7)	1.57(m)	1.54(m)
	0.93(t,7)		0.89(t,7)	0.92(m)
Propionic acid	2.42(q,7)	2.26(q,7)	2.47(q,7)	-
	1.18(t,7)	1.03(t,7)	1.10(t,7)	-
Pyridine	8.60(m)	8.61(m)	8.50(m)	8.53(m)
	7.69(m)	7.83(m)	7.90(m)	7.84(m)
	7.28(m)	7.40(m)	7.46(m)	7.43(m)
Succinimide	2.75	2.63	2.78	-
Tetrahydrofuran	3.74(m)	3.63(m)	3.75(m)	3.72(m)
	1.85(m)	1.78(m)	1.88(m)	1.87(m)

Note: The peaks listed are singlets, unless described as doublets (d), triplets (t), quartets (q) or multiplets (m). Coupling constants (in Hz) are given in parentheses.

5.2   Group 1 – Exchangeables and Aldehydes

Of all the protons you may encounter in an NMR spectrum, exchangeables (any protons that exist in a state of dynamic equilibrium with any free water that might be present in the solvent i.e. –OH, –NHR, –SH, –COOH, etc.) can be the least predictable – both with regard to their shape and their position. A guide to their typical chemical shift ranges and any notable features is given overleaf in Table 5.2. An alkyl –OH or –NHR for example, may be sharp and uncoupled, sharp and coupled, or broad and partially coupled. In a molecule with numerous exchangeables, they may appear distinct, or they may combine with each other, and with any water present – watch out for it particularly in D_6-DMSO solutions! Remember also that exchangeable protons will not be present in spectra of compounds run in D_4-MeOH, or D_2O solutions because they will have exchanged for deuterium. This forms the basis of a useful method for the identification of exchangeable protons which we will discuss in Chapter 7.

The origin of this unpredictability lies in the fact that they are relatively acidic and can undergo exchange in solution. The appearance of the signals we observe is governed by the rate at which this process occurs, the rate being greatly influenced by the nature of the solvent, its water content, pH, temperature and concentration of the compound.

Table 5.2   Typical exchangeable protons.

Exchangeable	Typical shift (ppm)	Remarks
Alkyl-OH	5–1	Can appear sharp and are capable of coupling to adjacent protons in dry aprotic solvents. Easily exchanged by shaking with D_2O.
Phenolic-OH	11–5	Often broad. Easily exchanged.
Phenolic-OH (H-bonded)	17–11	Can be broad but more usually sharp as proton exchange is slowed by need to break both bonds. Can therefore be more difficult exchange. Warm if necessary.
Alkyl-COOH	12–6	Usually broad but can be extremely broad! Very easily exchanged.
Aryl-COOH	14–8	As for Alkyl-COOH.
Alkyl-NH$_2$/NHR	5–1	Generally similar to alkyl-OH but maybe somewhat broader even in dry solutions and less likely to couple to adjacent protons. Ability to protonate nitrogen tends to broaden protons and displace to lower field. Easily exchanged.
Aryl-NH$_2$/NHR	11–6	Usually broad. Easily exchanged.
Alkyl-CONH$_2$/-CONHR	9–7	Often broad but frequently couple. Primary amides often appear as two broad signals due to partial double bond character of amide bond. Often slow to exchange and may require warming/mild base.
Aryl-CONH$_2$/-CONHR	13–7	As for Alkyl-CONH$_2$/-CONHR.
Alkyl-SH	5–1	Usually sharp and couple to adjacent protons. May need mild base to exchange (e.g. drop of $NaHCO_3/D_2O$ soln.) Beware easy oxidation to –S–S–. Therefore important to locate!
Aryl-SH	7–3	Somewhat broader and easier to exchange than alkyl-SH. Again, an important one to find!
Alkyl/aryl-SO$_3$H	14–6	Similar to corresponding –COOH.

A discussion of the kinetics of the process is outside the scope of this book because it won't help you to interpret your spectrum, but it is worth considering the two extremes of exchange, and the all-important region which lies between these extremes, as this might give you an insight into the seemingly fickle behaviour of exchangeable protons.

If you take a pure sample of ethanol, and run its NMR spectrum in dry $CDCl_3$, the hydroxyl proton will appear as a well-defined triplet, which couples to the adjacent $-CH_2-$, rendering it a multiplet. This is because the hydroxyl proton remains on the oxygen for relatively long periods of time, as there is nothing in the solution to entice it off, i.e. exchange (if any) is said to be very slow on the NMR timescale (slower than about 1 second).

The presence of a trace of acid and water, however, causes collapse of the hydroxyl-OH to a singlet (at lower field), the proton can now protonate, and de-protonate the oxygen very rapidly, as the process is catalysed by the acid i.e. exchange is said to be fast on the NMR timescale (faster than about 10^{-6} seconds).

In practice, one often encounters exchangeable protons which are exchanging at an intermediate rate, which leads to broadening of their signals, and only partial coupling (which can manifest itself as a mere broadening of the exchangeable proton and any they couple to). The actual position of the centre of a broad exchangeable signal, is dependent on how much water (or alternative exchange site) is present, and on the difference in the chemical shifts of the proton in the two environments. For example, a carboxylic acid proton, in a very dry solvent, may occur at about 12 ppm. A similar molar quantity of water in DMSO would absorb at around 3.5 ppm, and in such a solution, the carboxylic acid proton may well appear as a very broad signal, centred between these positions.

It is worth mentioning, that >NH protons may often appear somewhat broader than their –OH counterparts, for another reason – >NH protons have another relaxation mechanism available to them: quadrupole relaxation, because the ^{14}N nucleus has an electric quadrupole moment. This extra relaxation capability can lead to a shorter relaxation time for >NH protons, and since the natural linewidth of a peak is inversely proportional to the relaxation time of the proton(s) giving rise to it, a shorter relaxation time will give rise to a broader peak.

This can lead to cases where an >NH of an amide, for example, couples to a $-CH_2$-adjacent to it, without *appearing* to show a reciprocal coupling itself, which as we know, is impossible! What happens is that its coupling becomes lost in the broadness of the signal – consider the compound shown below and its spectrum (Spectrum 5.1).

Should your spectrum contain a very broad signal, such as the carboxylic acid proton of 4-fluoro benzoic acid shown in Spectrum 5.2 and you aren't sure whether it's there at all, or whether your eyes are deceiving you, try looking along the baseline. Any slight lump which could be a signal will be seen more easily in this way. Of course if you are operating the spectrometer yourself, you only have to turn up the vertical gain but if you are looking at a walk-up spectrum, this trick might be useful.

As we have mentioned already, a very useful tool when trying to identify exchangeables is to exchange them! – for deuterium, which removes them from the spectrum. This will be covered in detail in Chapter 7 but don't be in too much of a hurry to do this – they are part of the spectrum and hold valuable information. If they are sharp enough, they may become potentially useful targets for NOE experiments which we will discuss later.

Finally, a brief word about aldehydes. They are included at the end of this group for convenience only and should be spotted easily. Aldehydes bound to aromatic rings give sharp singlets at 10.2–9.9 ppm, while in alkyl systems, they give sharp signals at 10.0–9.7 ppm, which couple to adjacent alkyl protons with a relatively small coupling constants (2–4 Hz).

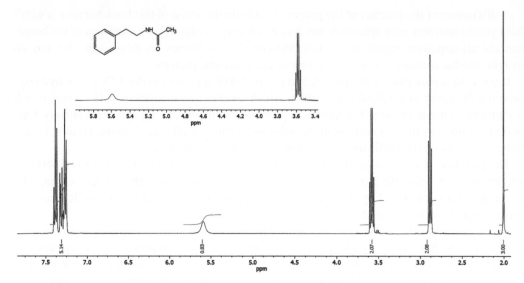

Spectrum 5.1 An amide NH (5.52 ppm) appearing to show no coupling to –CH₂ (3.51 ppm).

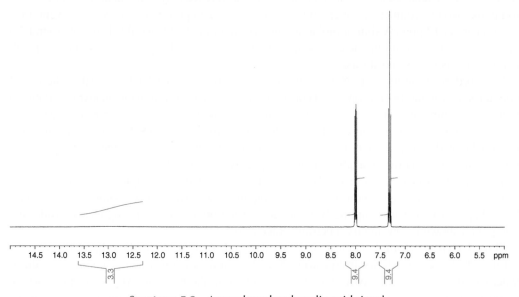

Spectrum 5.2 A very broad carboxylic acid signal.

5.3 Group 2 – Aromatic and Heterocyclic Protons

Protons on aromatic rings are generally fairly predictable, both as regards to their position and shape. The effects of substituents on a benzene ring are shown in Table 5.4.

They are applicable to compounds in the common NMR solvents – but not in D₆-benzene (or D₅-pyridine). The substituent effects are additive, but don't place too much reliance on chemical shifts predicted using the table, in compounds where more than two groups are substituted next to each other, as steric interactions between them can cause large deviations from expected values. Note that Table 5.4, like all others, does not cater for solvent shifts, etc.!

A number of features become apparent on running an eye over these figures. Firstly, one saturated carbon in a substituent between the benzene ring and another group (e.g. –CH₂–OH) is sufficient to virtually isolate the ring from the influence of that other group i.e. in this case, the –OH. This assumes that there are no abnormal 'through-space' effects, of course, which we'll touch on later.

Secondly, groups which withdraw electrons (e.g. $-NO_2$, –COR, –COOR) cause shifts of the aromatic protons to lower field, to varying extents around the ring (the ortho-protons are generally the most influenced by a substituent, followed by the para-protons, and the meta-protons being the least influenced). But some groups which are known to be electron-withdrawing in alkyl systems, such as –OH, –OR, –NR actually bring about upfield shifts in aromatic systems. This is because, while these groups withdraw electrons inductively, they more than make up for this by supplying electrons, mesomerically. This effect is almost exactly balanced in the case of –Cl, which has very little influence on aromatic protons.

As for spin coupling around the benzene ring, Table 5.3 shows the expected ranges and typical values:

Table 5.3   Aromatic proton–proton coupling constants.

Relative position	Coupling range	Typical coupling
Ortho-	6.0–9.4 Hz	8 Hz
Meta-	1.2–3.1 Hz	2.5 Hz
Para-	0.2–1.5 Hz	Negligible

Table 5.4   Aromatic protons – the common substituent effects.

| Substituent | Change in chemical shift in ppm relative to benzene (7.27) | | |
	Ortho	Meta	Para
$-NO_2$	0.95	0.17	0.33
–CHO	0.58	0.21	0.27
–COCl	0.83	0.16	0.30
–COOH	0.80	0.14	0.20
$-COOCH_3$	0.74	0.07	0.20
$-COCH_3$	0.64	0.09	0.30
–CN	0.27	0.11	0.30
–Ph	0.18	0.00	−0.08
$-CCl_3$	0.80	0.20	0.20
$-CHCl_2$	0.10	0.06	0.10
$-CH_2Cl$	0.00	0.01	0.00
$-CH_3$	−0.17	−0.09	−0.18
$-CH_2CH_3$	−0.15	−0.06	−0.18
$-CH(CH_3)_2$	−0.14	−0.09	−0.18
$-C(CH_3)_3$	0.01	−0.10	−0.10
$-CH_2OH$	−0.10	−0.10	−0.10
$-CH_2NH_2$	0.00	0.00	0.00
–F (couples!)	−0.30	−0.02	−0.22
–Cl	0.02	−0.06	−0.04
–Br	0.22	−0.13	−0.03
–I	0.40	−0.26	−0.03
$-OCH_3$	−0.43	−0.09	−0.37
$-OCOCH_3$	−0.21	−0.02	0.00
–OH	−0.50	−0.14	−0.40
$-NH_2$	−0.75	−0.24	−0.63
$-SCH_3$	−0.03	0.00	0.00
$-N(CH_3)_2$	−0.60	−0.10	−0.62

Note that a +ve sign denotes a *downfield* shift i.e. a shift to larger delta number (signal moves to the left).

Note that while in saturated systems, proton–proton couplings are seldom observed beyond three bonds, in aromatic and heterocyclic systems, 4- and even 5-bond coupling is commonplace. This is because spin coupling is transferred by electrons. Where you have extended conjugation, you can expect to observe coupling over a greater number of bonds.

In the light of this information, we can now consider a few examples of frequently encountered benzene-substitution patterns.

5.3.1   Monosubstituted Benzene Rings

Spectrum 5.3 shows a typical pattern for a benzene ring monosubstituted with a relatively 'electron neutral' group. In this case, it's plain old chloro-benzene but similar patterns can be expected for any relatively neutral substituents i.e. groups that neither donate nor withdraw electrons to or from the ring to any great extent (e.g. alkyl substituents). As the group has only a slight effect on the aromatic protons, they all resonate at quite close chemical shifts, giving anything from what can essentially be a singlet with small 'fringy bits' at its base, through to broader, heavily roofed multiplets as in this case though the exact appearance will of course vary considerably with spectrometer frequency. The complexity of the multiplet observed is dependent on two phenomena. The first, non-first order behaviour, we will discuss below. The second, magnetic non-equivalence, we will discuss in the following section which covers multi-substituted benzene rings.

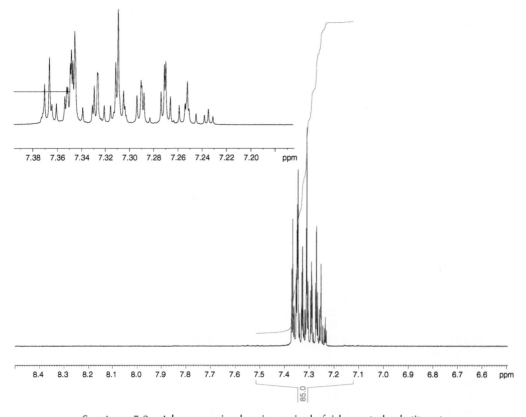

Spectrum 5.3   A benzene ring bearing a single fairly neutral substituent.

Splitting patterns of signals are nice and predictable, only as long as the protons coupling to each other are separated by a chemical shift which is large relative to the size of the coupling between them. Notice for example, how the triplet and quartet of an ethyl signal are almost perfectly symmetrical. However, when signals coupled to each other are much closer together in the spectrum so that the difference between their chemical shifts and the size of their coupling is comparable, non-first-order effects can be expected. The closer they are, the more distorted (more non-first order) they will be. In cases where signals are *very* close together, energy levels become mixed, and to quote L.M. Jackman, 'We find multiplicity rules no longer hold. Usually, more lines appear, and simple patterns of spacings and intensities are no longer found.' Such complex patterns can, in some cases, be subjected to mathematical analysis and the coupling information they contain extracted, but this practice has thankfully virtually died out with the advent of high-field spectrometers, or at least become a job for the computer! Our chlorobenzene example would look far less resolved (and thus more complex) than it does had it been run on an old 90 MHz instrument! Understanding, or at least recognizing 'non-first-orderness' is very important and relevant to interpreting the spectra you will encounter.

Spectrum 5.4 shows a typical pattern of a benzene ring monosubstituted with an electron-donating group (in this case, it's $-NH_2$).

Of the five aromatic protons, notice from the integration that two of them are below 7 ppm, occupying a position only slightly upfield of benzene itself, while the other three have been shifted upfield, above 7 ppm. A glance at Table 5.4 will show that these high-field signals can be assigned to the ortho- and para- protons. The meta- protons have been 'left behind', as it were. This shielding at the ortho- and para- positions is characteristic of simple electron-donating substituents.

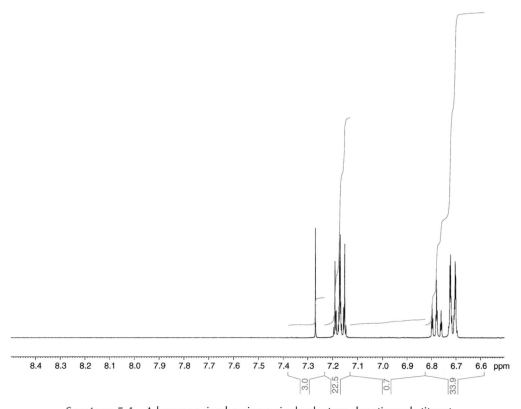

Spectrum 5.4   A benzene ring bearing a single electron-donating substituent.

These observations are of course underpinned by spin coupling observations. The meta- protons both experience two ortho- couplings of about 8 Hz which should yield a triplet (or more correctly, a doublet of doublets – note that a proton coupled to two other protons, which are different from each other, gives a doublet of doublets – when the two couplings are the same size, the signal appears as a triplet). What we actually observe is a very distorted 'triplet', the intensity of its lines being nothing like the 1:2:1 you might expect if you took Pascal's triangle too seriously! This distortion is referred to as 'roofing', and is the initial manifestation of the non-first-order behaviour just discussed.

The ortho- protons are shielded to the greatest extent and appear as a 'roofed' doublet of doublets while the less shielded para- proton presents as a triplet of triplets constructed from two large ortho- couplings and two small meta- ones. In cases where the electron-donating substituent is oxygen-based (i.e. –OH or –OR), para- shielding can be as large as the ortho-shielding so that the ortho- and para- protons may have very similar chemical shifts. The consequence of this will be explored further in the next section.

In Spectrum 5.5, we see the effect of a single deshielding substituent (carboxylic acid) on the benzene ring.

This time, we observe a pronounced downfield shift of the protons ortho- to the deshielding substituent and note that the signal is dominated by the large ortho-coupling and that it also bears a smaller meta- one. The signal is however both 'roofed' and is composed of more lines than you might naively expect…

The meta- and para- protons themselves appear as one ill-defined multiplet, but on closer inspection, you can see that they are just resolved from each other. The para- proton is slightly more deshielded than the meta- protons and is centred at 7.39 ppm and is in fact a heavily 'roofed' triplet of triplets.

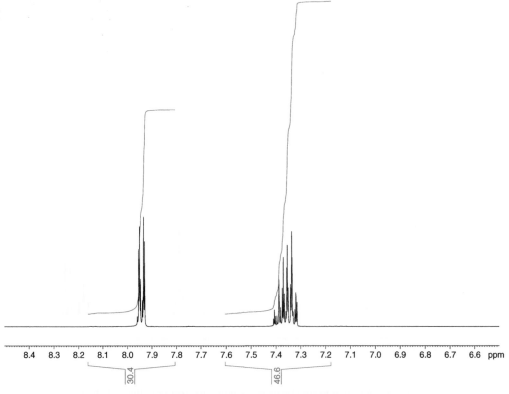

Spectrum 5.5 A benzene ring bearing a single electron-withdrawing substituent.

Note that Spectra 5.3, 5.4 and 5.5 are all plotted on the same scale to give you a feeling for the range of shifts that are typically encountered when looking at aromatic protons.

5.3.2 Multi-substituted Benzene Rings

Moving on to multi-substituted aromatic systems, the real value of Table 5.4 soon becomes apparent. In dealing with such systems, it will not be long before you encounter a 1,4 di-substituted benzene ring. This substitution pattern (along with the 1,2 symmetrically di-substituted systems) gives rise to an NMR phenomenon that merits some explanation – that of *chemical* and *magnetic* equivalence and the difference between them. Consider the 1,4 di-substituted aromatic compound shown below (4-bromobenzamide):

In terms of *chemical* equivalence, (or more accurately, *chemical shift* equivalence) clearly, H^a is equivalent to $H^{a'}$. But it is not *magnetically* equivalent to $H^{a'}$ because if it were, then the coupling between H^a and H^b would be the same as the coupling between $H^{a'}$ and H^b. Clearly, this cannot be the case since H^a is ortho to H^b but $H^{a'}$ is para to it. Such spin systems are referred to as AA'BB' systems (pronounced *A-A dashed B-B dashed*). The dashes are used to denote magnetic non-equivalence of the otherwise chemically equivalent protons. What this means in practice is that molecules of this type display a highly characteristic splitting pattern which would be described as a pair of doublets with a number of minor extra lines and some broadening at the base of the peaks (Spectrum 5.6).

These extra lines are often mistakenly thought to be impurity peaks. An in-depth understanding of how they may arise is not really necessary for the purpose of interpretation. What is important is that you instantly recognise the appearance of such spin systems. Check that the system integrates correctly and check that the two halves of the system are symmetrical. *Note: This phenomenon has nothing whatsoever to do with chiral centres and is purely a function of the spatial arrangement of the protons involved as described above.*

Spectrum 5.6 also shows a good example of 'roofing', which we touched on earlier. If you imagine the simple case of a pair of doublets well separated from each other, then all four of their lines will be of almost equal intensity. But when coupled doublets get closer together, they become distorted so that their inner lines become more intense, and their outer lines less intense. This is the onset of 'non-first-orderness'. The closer a pair of coupled doublets are to each

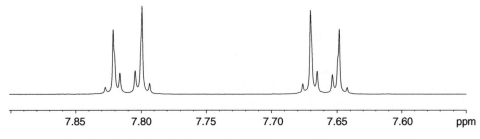

Spectrum 5.6   A typical aromatic AA'BB' system (4-bromobenzamide).

other, the more extreme the effect becomes. It is worth noting that the phenomenon can sometimes be a useful interpretive tool, as the roofing can indicate which doublet is coupled to which other one, in spectra where you encounter two or more systems of this type: doublets which are coupled to each other, always roof towards a point between them, as shown.

Obviously, there are too many possible combinations of groups for us to show a comprehensive collection of them but Spectrum 5.7 shows a nice example of a 1,3 di-substituted pattern featuring two strongly deshielding groups (a nitro group and a methyl ester) and serves to demonstrate the limitations of Table 5.4.

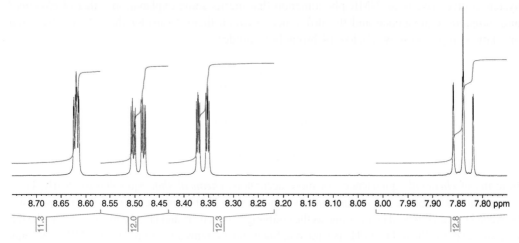

Spectrum 5.7   Methyl 3-nitrobenzoate.

Predicting the chemical shift of the proton between the two substituents using Table 5.4 suggests a figure of 8.96 ppm. The observed figure is in fact 8.62 ppm. Very low field for sure, but significantly not as low as predicted. We find that this sort of error is quite commonplace in ring systems containing two or more very deshielding groups. Naively, it's as if the first group withdraws so much electron density from the ring that there is not much left for the second group to withdraw so the combined effect is less than expected. Be that as it may, Table 5.4 at least succeeds in predicting the correct *relative* positions of the protons, even if the actual values are a little off the mark.

And finally, Spectrum 5.8 shows a classic example of a 1,2,4 tri-substituted benzene ring, (well-known anti-asthma drug, salbutamol) the structure of which is shown below… Obviously, the scope for variation in these systems is vast!

As a closing observation, it is difficult to say just how close you can reasonably expect predicted and observed values to be, even discounting highly sterically interactive systems mentioned earlier. A crude observation would be that the more substituents on the ring, the less accurate your predictions are likely to be. For what it's worth, however, a rough working guide would be an expectation of, shift predictions within 0.3 ppm for multi-substituted rings in the absence of strong steric interactions between groups.

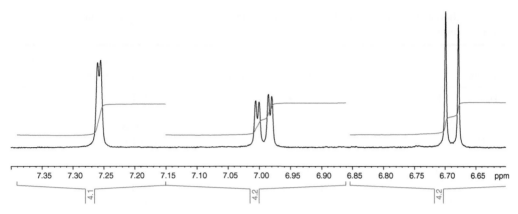

Spectrum 5.8   Salbutamol (aromatic region).

A final word of caution on aromatic systems – the electron-donating groups (notably those in which oxygen is the shielding entity) can cause problems, because their ortho- and para- effects are so similar. Consider the following example:

You are presented with a sample known to be one of the two isomers shown below:

The two compounds will give very similar spectra and you would not be able to tell which isomer your sample is without an authentic spectrum of at least one of the isomers, for comparison. The only unambiguous way to tell these isomers apart, in the absence of an authentic spectrum, would be by a nuclear Overhauser experiment (NOE), which we'll discuss later. Without performing such an experiment, you'd be ill-advised to chance your arm! Any chemical shift differences would be far too small to exploit with any certainty whatsoever!

5.3.3   Heterocyclic Ring Systems (Unsaturated) and Polycyclic Aromatic Systems

Heterocyclic systems resemble aromatic systems in some respects, but are more varied and interesting. We'll outline a few of these interesting features, and then provide some useful chemical shift and coupling data in Table 5.5. It is not really feasible to provide information as in Table 5.4, as every heterocycle would need its own specific table and there are a great many heterocycles out there!

So when confronted with a problem involving an unsaturated heterocycle, our advice is to make yourself aware of the shifts and couplings of the parent compound (Table 5.5) and then use the known effects of substituents from Table 5.4 and 'superimpose' them. This will give you a rough guide only and your 'Confidence Curve' will need to be adjusted accordingly as the magnitude of the induced shifts are usually somewhat different and may vary within the heterocycle. In a 3-substituted furan, or thiophene, for example, the magnitudes of the ortho-effects to the 2′, and 4′ protons are different, – sometimes, considerably so!

Table 5.5   Chemical shifts and couplings in some common heterocyclic and polycyclic aromatic systems.

Compound	Chemical shift (ppm)	Typical couplings (Hz) in parent *or* derivative
	a/d 7.4 b/c .3	a-b 1.8 b-c 3.5 a-c 0.8 a-d 1.6
	a/d 7.19 b/c 7.04	a-b 4.7 b-c 3.4 a-c 1.0 a-d 2.9
	a/d 6.62 b/c 6.05	a-b 2.6 b-c 3.4 a-c 1.1 a-d 2.2
	a/c* 7.55 b 6.25 * Note that tautomerism renders 'a' and 'c' equivalent in the parent NH compound.	a-b* 2.9 b-c *1.6 a-c 0*.7 * Couplings measured in non-tautomeric alkylated derivatives.
	a 7.7 b/c* 7.14 * Note that tautomerism renders 'b' and 'c' equivalent in the parent NH compound.	b-c 1.6 a-b ≈ a-c 0.8–1.5
	a 7.95 b 7.69 c 7.09	b-c 0.8 a-b 0.5 a-c 0.0
	a 8.88 b 7.41 c 7.98	b-c 3.1–3.6 a-b 1.9 a-c 0.0
	a/e 8.6 b/d 7.28 c 7.69	a-b 4–6 b-d 7–9 a-c 0–2.5 a-e 0–0.6 b-d 0.5–2 a-d 0–2.5

Table 5.5 (Continued)

Compound	Chemical shift (ppm)	Typical couplings (Hz) in parent *or* derivative
	a/d 9.17 b/c 7.68	a-b 5.0 b-c 8.4 a-c 2.0 a-d 3.5
	a/c 8.6 b 7.1 d 9.15	a-b 5.0 a-d 0 a-c 2.5 b-d 1.5
	a/b/c/d 8.5	a-b 1.8–2 a-d 0.5 a-c 1.5
	a/d 7.67 b/c 7.32	a-b 8–9 b-c 5–7 a-c 1–2 a-d ≈ 1 a-e ≈ 1
	a/d 7.98 b/c 7.44 e 8.40	Very similar to naphthalene above.
	a 8.65 b 7.61 c 7.57 d 7.86 e 7.70	a-b 8.4 b-c 7.2 e-f 9 a-c ≈ b-d 1.2 a-d ≈ 0.7

(continued)

Table 5.5   (Continued)

Compound	Chemical shift (ppm)	Typical couplings (Hz) in parent *or* derivative
	a 7.5 b 6.66 c 7.5 d 7.13 e 7.2 f 7.4	a-b 2.5 c-d 8 d-e 7.3 e-f 8.4 c-f 0.8 b-f ≈ 1
	a 7.26 b 6.45 c 7.55 d 7 e 7.1 f 7.4 g 9–12 (very solvent dependent!)	a-b 3 c-d ≈ e-f 8 d-e 7 c-e ≈ d-f 1.3 b-f 0.7 c-f ≈ 1 b-f 0.7 g-a 2.5 g-b 2
	a 7.44 b 7.34 c 7.83 d 7.36 e 7.34 f 7.9	a-b 5.5 c-d ≈ d-e ≈ e-f 7–8 c-e ≈ e-f ≈ 1 b-f 0.8

One other, perhaps even more dramatic and common example concerns compounds like 2′ and 4′ hydroxy- and amino-pyridines. These compounds exhibit tautomeric behaviour and tend to exist in solution as the corresponding pyridone and imine. This reduces the familiar pyridine-like properties of the ring system, accentuating the effects of these substituents (in terms of induced chemical shifts) and at the same time, radically increasing the expected couplings 2′-3′ couplings.

The size of couplings around heterocyclic rings can also vary dramatically. Ortho- couplings in five-membered heterocycles such as furan and thiophene, for example, are much smaller than in normal aromatic rings. Note also that even within a given heterocycle, there can be substantial variation in the size of ortho couplings themselves! As with any spectroscopic phenomenon, this should not be regarded as just another complication, but as an important part of your spectroscopic armour, or indeed, as part of your spectroscopic offensive weaponry for attacking problems of substitution, etc.

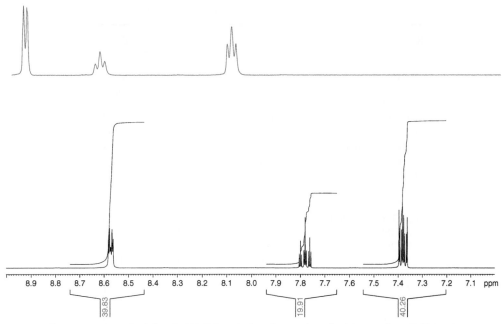

Spectrum 5.9 Pyridine in DMSO solution (bottom) and with one drop DCl (top).

Nitrogen-containing heterocycles are sometimes basic enough to protonate and form salts in acidic conditions and this leads to substantial changes in chemical shifts of their protons – see Spectrum 5.9 (pyridine and pyridine + DCl).

Note also that fluorine couplings to protons in heterocyclic systems can be well outside intuitive expectations! See Section 11.8 on $^{19}F–^1H$ coupling for an example!

One last word on heterocycles. Very small couplings (<1 Hz) have been found to exist between some protons on *different* rings of bicyclic heterocycles. For example, in indole, there is a 3–7 coupling of about 0.7 Hz. In practice, however, these very small couplings may only manifest themselves as a broadening of the signals concerned.

Obviously, Table 5.5 is far from exhaustive but it establishes the typical shifts and couplings found in some of the more commonly encountered heterocycles.

5.4 Group 3 – Double and Triple Bonds

In this section, we will look at alkene, imine, enol-ether and alkyne protons. It's convenient to consider the first three at this stage as they usually show signals in the 8-5 delta region and the alkyne is included here for convenience.

Alkene chemical shifts can be estimated using Table 5.6. Use this table with the same circumspection as you would all other tables of this type. It's a useful guide, not gospel.

Substitute the additive values in Table 5.6 into the following equation:

$$\text{Approximate chemical shift of proton (ppm)} = 5.25 + Z_{\text{gem}} + Z_{\text{cis}} + Z_{\text{trans}}$$

Table 5.6   Estimation of chemical shifts for alkene protons.

R	Z_{gem} (ppm)	Z_{cis}	Z_{trans}
–H	0.00	0.00	0.00
–Alkyl	0.45	−0.22	−0.28
–CH_2–OR	0.64	−0.01	−0.02
–CH_2–SR	0.71	−0.13	−0.22
–CH_2–halogen	0.70	0.11	−0.04
–CH_2NR_2	0.58	−0.10	−0.08
>C=C< (isolated)	1.00	−0.09	−0.23
>C=C< (conjugated)	1.24	0.02	−0.05
–CN	0.27	0.75	0.55
–C $\equiv$ C-R	0.47	0.38	0.12
>C=O (isolated)	1.10	1.12	0.87
>C=O (conjugated)	1.06	0.91	0.74
–COOH	0.97	1.41	0.71
–COOR	0.80	1.18	0.55
–CHO	1.02	0.95	1.17
–$CONR_2$	1.37	0.98	0.46
–COCl	1.11	1.46	1.01
–OR	1.22	−1.07	−1.21
–OCOR	2.11	−0.35	−0.64
–CH_2–Ar	1.05	−0.29	−0.32
–Cl	1.08	0.18	0.13
–Br	1.07	0.45	0.55
–I	1.14	0.81	0.88
–NR_2	0.80	−1.26	−1.21
–NRCOR	2.08	−0.57	−0.72
–Ar	1.38	0.36	−0.07
–SR	1.11	−0.29	−0.13
–SO_2R	1.55	1.16	0.93

Typical couplings found in alkenes are shown below:

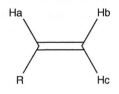

J_{a-b} (cis coupling)	7–13 Hz, typically 11 Hz
J_{a-c} (trans coupling)	11–18 Hz, typically 16 Hz
J_{b-c} (geminal coupling)	approx. 1–2 Hz

These couplings are exemplified below with reference to styrene (Spectrum 5.10).

Note that small couplings (1–2.5 Hz approx.) would also be expected between the first CH_2 of any alkyl group (R) and both Hb and Hc.

The actual sizes of the observed cis- and trans- couplings are influenced by the electronegativity of the substituents attached to the double bond. In general, the more electronegative the substituents,

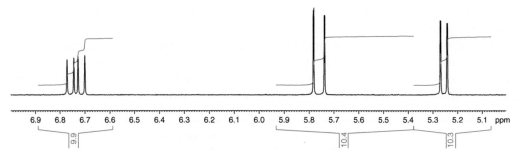

Spectrum 5.10 The alkene protons of styrene (Ph–CH=CH$_2$).

the smaller the observed couplings. (There is actually an approximately linear relationship between the size of the coupling and the sum of the electronegativities of the substituents.)

It is interesting to note that in cases where an alkene is joined directly to an aromatic ring, the alkene proton geminal to the aromatic ring is invariably at the lowest field of the alkene protons. This is because the alkene bond tends to lie in the same plane as the aryl ring and for this reason, the geminal proton is held in the deshielding zone of the aromatic ring, as is the alkene proton cis to the aromatic ring. This is an example of anisotropy which we will discuss in some detail later on.

Determining whether an alkene is cis or trans in cases where the alkene is in the middle of a alkyl chain is usually not possible by ^{1}H NMR as both cis and trans protons have very similar shifts in such circumstances, as do the –CH$_2$s attached to the alkene. Such a problem can however be dealt with using ^{13}C NMR where the shifts of these CH$_2$s are diagnostic.

Other double-bond moieties which are often encountered include the imines and oximes.

Imine   Oxime

Note that the lack of rotation about the double bond means that 'E' and 'Z' isomers are distinct entities in the same way that cis and trans isomers are distinct in conventional alkenes. It is not really feasible to give a comprehensive guide to the chemical shifts of these protons but expect them to be somewhat lower-field (approx. 1 ppm) than for comparable alkenes, with chemical shifts being driven largely by the anisotropy of the substituents.

Enol-ether protons are interesting in that their chemical shifts are unusually high field in comparison with other alkenes on account of lone-pair donation into the double bond from oxygen. No special precautions are necessary when dealing with them as this is reflected in the values obtained using Table 5.6.

(4.2 δ) H   OR (R = long-chain hydrocarbon)

(3.9 δ !) H   H (6.5 δ)

An example of an enol-ether showing typical shifts.

Whereas alkene protons are relatively deshielded by the overlapping p-electrons of the double bond, alkyne protons are fairly shielded by their electronic environment. In common with alkenes, however, is the possibility of small, long-range coupling through the triple bond. The chemical shifts of alkyne protons are highly influenced by the electronegativity of groups attached to the other end of the triple bond as can be seen from the examples below. It is worth bearing in mind that alkyne protons may exchange in strongly basic solutions.

Alkyne	Chem. Shift (ppm)	Comments
R–C≡C–**H**	1.9	(R=long-chain hydrocarbon)
Ph–C≡C–**H**	3.1	
Ph–Ph–C≡C–**H**	4.2	(Ph–Ph=biphenyl rings)

5.5   Group 4 – Alkyl Protons

This section must necessarily be brief and general on account of the size of the category, and the vast number of case studies we could dissect in detail.

For now, the discussion will be restricted to straightforward systems (open-chain and containing no chiral centres) and adhere to previous practice by supplying chemical shift data (Table 5.7) which will enable you to estimate the chemical shifts of methyl, methylene and methine protons you will typically encounter. Typical 3-bond couplings in such systems can be expected in the region of 7–9 Hz, what variations there are being attributable to electronic

Table 5.7   Estimation of chemical shifts for alkyl protons.

X	C	X	C
–CH_3	0.5	–NR,COR	2.4
–Alkyl	0.6	–NR_2	1.6
≥<	1.3	–$^+NR_3$	2.4
–C≡C–Ar	1.7	–N=C=S	2.9
–C≡C–R	1.4	–Ar	1.8
–CN	1.7	–OCOAr	3.5
–COAr	1.9	–OCOR	3.1
–COCl	1.8	–OAr	3.2
–$CONR_2$	1.6	–OH	2.6
–COOR	1.5	–OR	2.4
–COR	1.6	–OSO_2Ar	3.4
–Cl	2.5	–SAr	2.1
–Br	2.3	–SR	1.9
–I	2.1	–CF_3*	1.1
–NO_2	3.7	–F*	3.6

Note: Both show coupling to neighbouring alkyl protons.
For methine protons, approx. chemical shift (ppm) will be 0.1 + CX + CX1 + CX2
For methylene protons, approx. chemical shift will be 0.3 + CX + CX1
For methyl protons, approx. chemical shift will be 0.5 + CX
Note: The values 0.1, 0.3 and 0.5 are just 'fudge factors' to give better estimates.
For more extensive shift and coupling data on a wider variety of compounds, we would recommend *Structure Determination of Organic Compounds* by
E. Pretsch, P. Bühlmann and C. Affolter (Springer, ISBN 3–540-67815–8).

effects of the substituents. These small variations can sometimes be exploited as a means of verifying which signal is coupled to which other e.g. in cases where you are up against a molecule with two different $-CH_2CH_2-$ systems. Perhaps we should mention at this stage, that the single most significant factor in determining the magnitude of a 3-bond (vicinal) coupling, is the dihedral angle φ, between the protons in question.

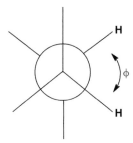

In open-chain compounds that lack any chiral centre of course, rotation about all single bonds can be assumed to be both relatively 'free' and fast on the NMR timescale and the 7–9 Hz range quoted is the result of averaging of this angle. The same is of course not true in cyclic systems where structures are rigid and bond angles constrained. We will deal with this topic thoroughly in Section 6.5.

Let's stop for a moment and reflect on what we have dealt with so far. In fact, we've covered quite a lot of ground already. We started by considering some basic theory and background to the subject. We've looked at the very important issues of sample preparation and skimmed the surface of spectrometer set-up. We've established a good standard method of dealing with spectra, by partitioning the information available into coherent segments – both with respect to the nature of the information (chemical shift, multiplicity and integration) and also, with respect to the various classes of proton commonly encountered. And finally, we've spent a good deal of time examining these different types of protons in some detail. In fact, it might be tempting to wonder what more needs to be said on the subject of spectral interpretation. After all, you now have in your grasp some pretty powerful tools – tables and so on, which will, if used prudently, give you a good idea of what to expect from relatively simple spectra.

Unfortunately, it's not quite as simple as that. In the next chapter, we'll delve a little deeper and have a look at some possible pitfalls you may encounter in more complex spectra…

6

Delving Deeper

6.1 Chiral Centres

It's fair to say, that if all molecules were flat and lacked chiral centres, the interpretation of their NMR spectra would be far easier than it actually is but it would be a whole lot less fun too! In moving on to discuss more interesting chiral compounds, we have an opportunity to deal with some commonly held misconceptions and urban myths that can severely limit understanding of the subject.

A good working knowledge of stereochemistry is certainly a big advantage when looking at the spectra of chiral molecules. Let's start by considering the molecule below:

Clearly, the bold-highlighted carbon is a chiral centre (it has four different groups attached to it). For this reason, the two protons H_a and H_b can never be in the same environment. The fact that there is free rotation around all the single bonds in the molecule is irrelevant. This can best be appreciated by building a model of the molecule. Having done so, look down the molecule from left to right as drawn and rotate the C–O bonds so that H_a and H_b rotate. It should now be clear why these two protons can never occupy the same space and are therefore not equivalent.

Now for the next big step forward… if they are not equivalent, then there is no reason for them to have the same chemical shift. Another big step… and if they have different chemical shifts, they will couple to each other… In fact, in molecules of this type i.e. that have an isolated CH_2 in the region of a chiral centre, the likelihood is that the CH_2 will be observed as a pair of doublets. (See Spectrum 6.1.)

How close they are to each other, or how far apart, is not something that can be easily estimated as it depends on the through-space interactions (anisotropies) of both protons with all the other groups in the molecule. That said, the two doublets are likely to be within 1 ppm of each other and are therefore likely to be clearly 'roofed' to each other. Spectroscopists use the term

Essential Practical NMR for Organic Chemistry, Second Edition. S.A. Richards and J.C. Hollerton.
© 2023 John Wiley & Sons Ltd. Published 2023 by John Wiley & Sons Ltd.

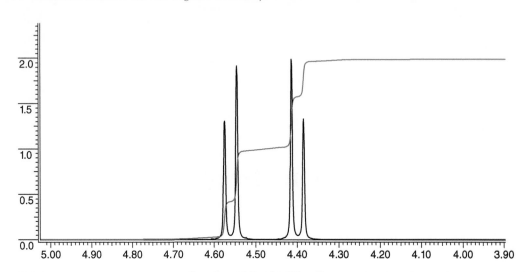

Spectrum 6.1   An AB system.

'AB system' to describe this type of arrangement. All it means is that the spin system contains two protons which are relatively close to each other in chemical shift terms (but not equivalent to each other), HA and HB, and they couple to each other and nothing else.

Geminal couplings of this type are typically in the region of 12–14 Hz, though interestingly, they can be as large as 19 Hz between protons that are alpha to an alkene or carbonyl function. This can be a useful interpretive 'handle' if you are looking for a starting point in a complex assignment. (Take a look at a spectrum of camphor if you need convincing! Spot any huge geminal couplings?)

Of course, it is quite possible, though statistically unlikely, that you might encounter a molecule of this type in which the chemical shifts of the two protons, H_a and H_b, just happen to be identical. Under these circumstances, there will be no splitting observed and you will just observe a singlet as if there were no chiral centre in the molecule at all. But beware! Should you run the sample in a different solvent, or even in the same solvent but at a different concentration, the singlet would be likely to re-present itself as an AB system. *Note: The degree of separation between H_a and H_b reflects the anisotropic influences the different groups on the chiral centre exert on the two protons. If these groups were all very similar in nature (e.g. an ethyl, propyl and butyl) there would be very little 'difference' engendered in H_a and H_b, and for this reason, we could reasonably expect the chemical shift difference between these two protons to be small.*

You might consider there to be an issue in predicting the chemical shift of a signal that is split into an AB system in this way but in reality, we have found it safe to treat the prediction as the mid-point between the two doublets of the AB.

So in summary, the presence of a chiral centre in a molecule can render nearby geminal pairs of protons non-equivalent. 'Nearby' is not an exact term and varies according to circumstance. Let's consider our molecule again, but this time, replace the –CH$_2$– with an alkyl chain…

In this case, it should be clear that H_a and H_b are just as non-equivalent as before. And because they are non-equivalent, it stands to reason that the next pair of protons, H_c and H_d must also be non-equivalent…and the next pair and the next pair… And so it is. In terms of the spectral lines observed, complexity will certainly be the name of the game! Not only will H_a and H_b couple to each other but they will obviously both couple to H_c and H_d. What will not necessarily be so obvious is that the size of the splittings between H_a and H_c and between H_a and H_d will very likely be different! This is because although there is free rotation about all single bonds, the chiral centre will place certain steric constraints upon the molecule such that it will tend to adopt a conformation that will minimise these constraints. This means that the time-averaged dihedral angles between H_a and H_c and H_a and H_d will not be the same – and neither will be the corresponding couplings. All of a sudden, in this welter of complex, overlapped, heavily roofed multiplets, Pascal's triangle starts to look woefully inadequate, doesn't it?

In practice, of course, we find that the further away from the chiral centre we go, the smaller the difference in chemical shift between corresponding geminal protons is likely to be. By the time we move three or four carbons down the chain, the likelihood is that corresponding pairs will be approximately equivalent, so for example, in the case above, we might expect the $-CH_2-$ next to the phenyl ring to be just a fairly normal, slightly broadened, roofed triplet rather than a pair of complex multiplets. It is not impossible, however, for a molecule to wrap itself up in certain conditions such that a geminal pair of protons are brought near to a chiral centre in the molecule – even though they may be many, many bonds away from it. It is important to remember that this is a 'through-space' effect rather than a 'through-bond' effect.

The convention of appending letters of the alphabet to protons in order to describe spin systems is commonly used in two more important cases. A molecule which would be likely to exhibit a classic ABX system is shown below. (See Spectrum 6.2.)

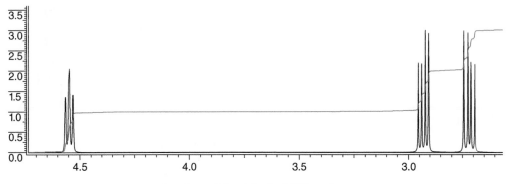

As before, the chiral centre renders H_a and H_b non-equivalent and for the reasons already covered, H_x will couple to both with all three couplings (H_a–H_b, H_a–H_x and H_b–H_x) likely to be different. So the classical presentation of an ABX system is that of three multiplets, each of four lines. (Note that in Spectrum 6.2, the size of the H_a–X and the H_b–X couplings are almost iden-

Spectrum 6.2   A typical ABX system.

tical so the X proton appears as an approximate triplet. This is quite common.) The AB part indicates that the geminal pair are likely to be relatively close in terms of chemical shift, while the X proton is someway distant from both. Obviously, the scope for variation in the appearance of ABX systems is enormous. The difference in chemical shift between H_a and H_b is a major factor in this but we have also come across ABX systems constrained within five-membered rings where all three splittings happen to be the same size. In such cases, we observe three triplets. Another possibility is for A and B to be accidentally equivalent in which case we observe something approximating to a simple doublet for H_a and H_b and a triplet for H_x.

It is also quite common to see molecules in which the X proton is actually the X of two distinct ABX systems. An example of such a molecule is shown below along with its spectrum (Spectrum 6.3).

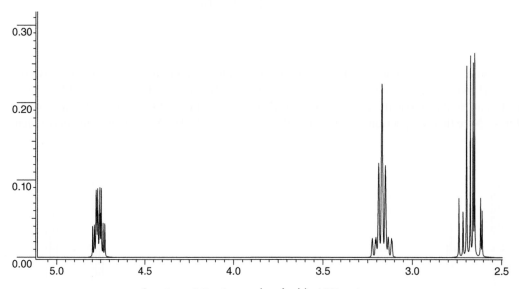

In a molecule like this, it would be theoretically possible for H_x to present as a 16-line multiplet but it is extremely unlikely that you would be able to count this many as there would almost certainly be a considerable overlap between them. Then of course, it is would be quite possible for the two AB parts to overlap... Be flexible in your approach and alert to the possibilities...

Moving on to some wider stereochemical considerations, just as enantiomers are indistinguishable as far as their physical and chemical properties are concerned (except, of course, as regards their reactions with other optically active reagents) so their spectra, acquired under normal conditions, are identical. The NMR spectrometer does not differentiate between

Spectrum 6.3   A complex double ABX system.

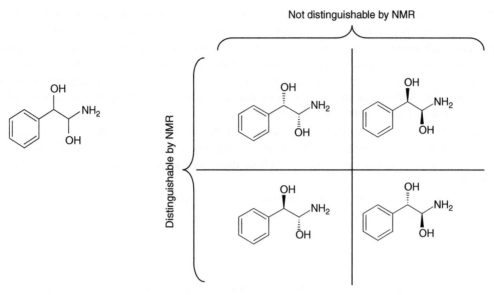

Figure 6.1 NMR and the relationship between enantiomers and diastereoisomers.

optically pure samples and racemic ones. *Note: There is a way of differentiating between enantiomers by NMR but it involves using certain chiral reagents which we'll discuss in detail later.*

So much for one chiral centre. The problems really begin when you come up against molecules which have two or more chiral centres. With two chiral centres, we can construct four possible stereoisomers. These can be separated into two enantiomeric pairs (indistinguishable by NMR). But, (KEY SENTENCE COMING UP) if we compare one member of each of these enantiomeric pairs, we will find that they may be distinguished from each other by NMR, because they are diastereoisomers. Diastereoisomers are stereoisomers which are not mirror images of each other – they are different compounds with distinct physical and chemical properties. See Figure 6.1 if this isn't clear.

Differences in the spectra of diastereoisomers are generally most noticeable in the region of the chiral centres. Spectrum 6.4 shows a typical example.

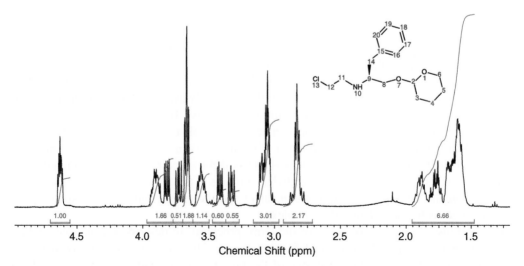

Spectrum 6.4 A mixture of diastereoisomers.

Note how two sets of signals are clearly visible, for the protons labelled '8' above. These present as two pairs of protons, i.e. two AB- parts of two ABX systems at 3.30–3.45 and 3.70–3.85 ppm, each integrating for approximately half a proton with respect to the unresolved parts of the spectrum. You certainly wouldn't expect *all* the signals of a pair of diastereoisomers to resolve (e.g. protons 3, 4 and 5 in the example above) but some will almost certainly do so. In some cases, the differences in the spectra of diastereoisomers can be quite spectacular, with chemical shift differences of 0.5 ppm or more.

With more than two unspecified chiral centres, problems multiply rapidly – three chiral centres yield eight stereoisomers, and thus four possible sets of signals and so on. From this, it follows that n chiral centres give rise to 2^n chiral entities of which $2^n/2$ will be distinguishable by NMR.

A final point on this phenomenon – nitrogen can sometimes act as a chiral centre. This topic is explored in some detail in Section 6.8.

6.2   Enantiotopic and Diastereotopic Protons

Consider ethanol (KEY SENTENCE COMING UP). If you were to replace each of the methylene protons in turn with some other group, Z, you would end up with a pair of enantiomers. We call this 'The Z-Test'. For this reason, the protons (or whatever groups may be involved, in molecules of the type: X-CA$_2$-Y where A can be either H or another group) are described as *enantiotopic*. This is of no consequence in the spectrometer, because as we have mentioned, enantiomers are not distinguishable by NMR under normal conditions.

So far so good. Now consider the following molecule:

The molecule clearly does not contain any chiral centres and so should give a perfectly straightforward spectrum. Now take a look at Spectrum 6.5.

On close examination, it is clear that the methylene protons of the –OEt groups (H4 and H6) do not give the nice simple quartet which we might reasonably expect. Close examination of the methylene signal shows it to be a complex multiplet. But why? Try applying the 'Z-Test' to the methylene protons. Straight away, the difference between this molecule and ethanol becomes apparent. Whereas ethanol would yield a pair of enantiomers in response to the test, this molecule would yield a pair of diastereoisomers as a second chiral centre would be generated at the branch point (C2)! For this reason, the methylene protons in this molecule would be described as *diastereotopic*. Such protons are not equivalent and therefore exhibit further splittings as they couple to each other – hence the complexity.

Some confusion can arise over use of the term 'prochiral' to describe various sites within molecules and is perhaps best avoided for this reason. The term means, literally, one step removed from being chiral (i.e. swap one of the protons for 'Z' and you have a full chiral

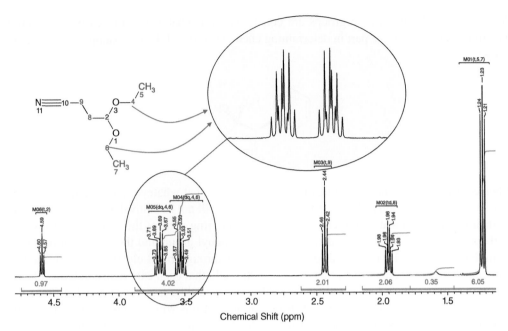

Spectrum 6.5 Diastereotopic protons.

centre). The methylene in ethanol for example, would be a good example. What we have in the di-ethoxy molecule above is one prochiral centre acting in combination with another to render a pair of protons non-equivalent.

6.3 Molecular Anisotropy

There are two factors that determine chemical shifts – electron distribution and molecular anisotropy. We have already seen how electronics define chemical shifts in previous sections. When we use Table 5.4 to estimate shifts around an aromatic ring, for example, the predictions we arrive at are based on the known electron withdrawal or supply of the various substituents on the ring. No allowance is made for unusual anisotropy. Similarly, predictions of chemical shifts of alkyl protons using Table 5.7 will be calculated on the basis of electronic factors only, as it would be impossible to vector anisotropy into the prediction since it varies in each individual molecule. They will be reasonably accurate in molecules where electronic factors predominate and molecular anisotropy has little or no influence. A typical example of such a molecule is shown below. Note the lack of steric crowding in the structure.

However, in molecules where groups are constrained for whatever steric reasons, molecular anisotropy can play a large part in determining chemical shifts. Take for example, the molecule below:

When confronted by a molecule like this, we can be sure that whatever conformation it adopts in solution, the likelihood is that the two methyl groups will not be equivalent. The driving force for their non-equivalence will of course be the aromatic ring. One of the methyl groups will be on the same face of the five-membered ring as the phenyl group and the other will not (once again, building a model is a good idea). In terms of through-bond electronics, both methyls enjoy much the same environment but the magnetic field that each experience in terms of their proximities to the phenyl ring will be very different. And this, in essence, is what molecular anisotropy is all about – non-uniform distribution of electrons within groups, inducing significant chemical shift changes in parts of molecules by the introduction of localised magnetic fields. In the molecule above, for example, it would be likely that the phenyl ring would interact with the methyl group cis to it (i.e. on same face of the five-membered ring) in such a way as to minimise contact. In order to do this, it would probably spend most of its time at right-angles to the plane of the paper i.e. sticking up vertically out of the page. This would position the cis methyl over the phenyl ring's pi-cloud which would induce an upfield shift in this methyl and cause it to be higher field than you might expect. The trans methyl would be relatively unaffected.

A most extreme example of anisotropy is demonstrated in the molecule in Figure 6.2.

In this unusual metacyclophane, the predicted chemical shift (Table 5.8) of the methine proton that is suspended above the aromatic ring would be 1.9 ppm. In fact, the observed shift is −4 ppm i.e. 4 ppm *above* TMS! The discrepancy between these values is all down to the anisotropic effect of the benzene ring and the fact that the proton in question is held very close to the delocalised 'p' electrons of the pi-cloud.

All groups have a certain measure of anisotropy associated with them so that any protons forced abnormally close to *any* group are likely to exhibit some deviation from expected chemical shifts but the most notable are the aromatic/heterocyclic groups, carbonyls and alkenes. Expect abnormal shifts in molecules where steric crowding forces groups into close contact

Figure 6.2   Anisotropy of a metacyclophane.

with each other. Build models and try to envisage the likely (lowest energy) conformations of your molecules. How will various groups within your molecules align themselves with respect to the anisotropic moieties? Remember that aromatic/heterocyclic rings shield groups that are held *above* or *below* their plane but deshield groups that are held *in* their plane and that groups held near the 'oxygen end' of a carbonyl group will be deshielded.

It is anisotropy that is the ultimate cause of the chemical shift differences between the geminal protons in AB and ABX systems. And indirectly, it is changes in anisotropy that bring about differences in observed chemical shifts for the same sample that is run in different solvents. The unknown extent of the anisotropy term in defining chemical shifts make it difficult (or perhaps impossible?) to devise a prediction tool, computer-based or otherwise, that can accurately predict the shifts of all protons regardless of environment. You may be wondering why the extent of the anisotropy term should be unknown. This is because in order to calculate it, we would first need to know the exact shape of the molecule in question – in solution. Molecular modelling packages deal with single molecules in a vacuum. This is nothing like variable concentrations in a variety of organic solvents with varying water content!

Molecular anisotropy affects proton chemical shifts to a far greater extent than ^{13}C chemical shifts. This is because the protons occupy the outer extremities of a molecule while the carbon framework is far more internal and, to a large extent, removed from the influences of anisotropy.

Always be aware of anisotropy but as with all NMR phenomena, avoid becoming obsessed with it!

6.4 Accidental Equivalence

Accidental equivalence is a fairly self-explanatory term used to describe situations where different signals happen to be coincidental. In most cases, this will come as no great surprise, and should cause no great problem. As you run through the mental exercise of estimating the chemical shifts of all the protons in your spectrum in order to create a hypothetical spectrum, don't forget to consider the possibilities of signals sharing exactly the same chemical shifts. There is no mysterious force acting to 'repel' chemical shifts away from each other and it is quite possible for chemical shifts to be coincidental. Take for example, the simple molecule of the type shown below:

Checking the chemical shifts for the alkyl chain *as a free base* would lead us to conclude that the $-CH_2-$ next to the aromatic ring should be at a slightly lower field (approx. 2.9 ppm, making allowance for the short chain length and slight beta deshielding effect of the $-NH_2$ group) than the $-CH_2-$ next to the $-NH_2$ function (should be approx. 2.6 ppm). However, should the nitrogen be protonated by an acid, then the $-CH_2-$ next to the $-NH_3^+$ would have the lower chemical shift (approx. 3.4 ppm) and those next to the aromatic ring, approx. 3.1 ppm (due to enhanced beta deshielding). The act of protonation causes the shift of one $-CH_2-$ to 'overtake' the other, as they both move downfield. But at a certain intermediate acid concentration, both $-CH_2-$s will have exactly the same chemical shift and will present as a four-proton singlet. If confronted with a situation like this, your first thought might be: 'This cannot be right!' But your second thought

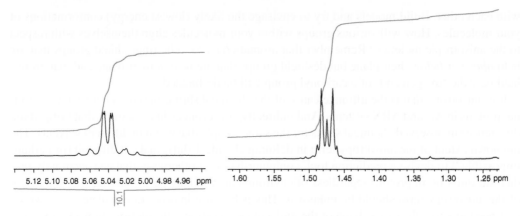

Spectrum 6.6   'Virtual coupling'.

should be: 'Ah! Maybe the nitrogen has been partially protonated by exposure to some acid?' And your third thought should be: 'Okay. So what am I going to do to prove this?'

As we have already seen, accidental equivalence could be responsible for the theoretically non-equivalent protons of an AB system presenting as a singlet and for the more complex ABX system presenting as a simple doublet and triplet. But occasionally, even more interesting manifestation of accidental equivalence can be observed. Consider the molecule below and its spectrum (Spectrum 6.6) which shows only the regions of interest to us – expanded and with the intervening region removed.

The complex multiplet centred at 5.04 ppm results from the overlap of the methine and –OH protons (i.e. they are 'accidentally equivalent') while the equally complex methyl signal is centred at 1.48 ppm. Because of this overlap, their lines are indistinguishable and so the –OH is said to be 'virtually coupled' to the methyl group. Virtual coupling is another potential consequence of non-first-order behaviour.

$$\text{HETEROCYCLE}{-}\overset{\displaystyle \text{OH}}{\underset{\displaystyle \text{CH}_3}{\big\langle}}$$

And for a final example, consider this molecule… (Spectrum 6.7). Please note: Spectrum 6.7 has been simulated on account of no compound being available at the time of writing. The chemical shifts and splitting values were taken from an actual spectrum published in *Laboratory Guide to Proton NMR Spectroscopy* (see Introduction).

When we look at this 'deceptively simple' spectrum, it soon becomes clear that two of the aromatic protons must be isochronous since we see only two multiplets with the appropriate integration of 2:1 for the three protons. The lowest field of the aromatic protons must be 'H_a' as it is ortho to the deshielding aldehyde function and therefore it must be the slightly higher-field protons 'H_b' and 'H_c' which are accidentally equivalent to each other as they are either ortho or para to the electron-donating (upfield-shifting) oxygen atoms. Were it not for the fact that 'H_b' and 'H_c' share the same chemical shift, we would expect to see them couple to 'H_a' with couplings of about 7.5 and 2.5 Hz, respectively. What we see in reality is an approximate triplet/doublet structure with an apparent splitting of about 5 Hz! This is clearly too large to be a meta coupling and too small to be an ortho coupling. Note that the small additional lines flanking the doublet and triplet

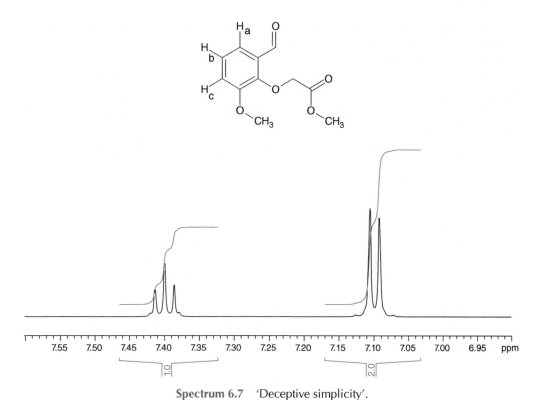

Spectrum 6.7 'Deceptive simplicity'.

are real and part of the signals in question. They can be explained by the magnetic non-equivalence of 'H$_b$' and 'H$_c$' and are a manifestation of non-first-order behaviour.

It is in effect a hybrid splitting; literally, an average of the two expected couplings. The two protons become indistinguishable from each other and both appear to exist in some hybrid ortho/meta state! The term 'deceptive simplicity' is quite apt to describe such a spin system. It might look simple, but it isn't. It's non-first-order splitting at its most beguiling!

Don't bother trying to find this sort of thing in your spectra. It is a rare phenomenon (and the more powerful your magnet, the rarer it is) and you won't find it. But it's good to be aware of it because if you look at enough spectra, one day *it* might find *you*...

6.5 Restricted Rotation

Certain types of bond, while nominally being considered as 'single', have in fact, sufficient 'double bond character', to render rotation about their axis, 'restricted'. The one you are most likely to encounter is the amide bond. Partial double bond character exists between the carbonyl, and the nitrogen, and may be represented thus:

This can lead to problems in NMR spectra. The magnitude of the energy barrier to the rotation determines what the effect on the spectrum will be. (For the thermodynamically-minded, we are talking about energy barriers of the order of 9–20 Kcal/mole.)

Should the energy barrier be substantially lower than this, then restriction will be slight, and rotation will be relatively fast on the NMR timescale, and therefore we may only see a slight broadening of signals in the region of the site of restricted rotation. Conversely, should the energy barrier be relatively high, rotation will be slow enough for us to see two distinct sets of signals. The worse-case scenario is that of rotation which is of intermediate pace on the NMR timescale, as this gives rise to broad semi-coalesced signals that are impossible to interpret.

Let's return to our amides. In primary amides, where R′ and R″ are both just protons, we can expect to see them as two, distinct, broad signals (Spectrum 6.8).

This is because the two protons do not occupy the same environment. Though they do exchange their positions with each other, the process is 'slow on the NMR timescale'. This means that during the time in which a single transient is acquired, there will have been relatively little exchange and, for this reason, the spectrometer will 'see' the two amide protons in two distinct environments and you will observe two distinct broad humps separated typically by about 0.6 ppm. Anisotropy of the carbonyl group ensures that the lower-field of the two humps corresponds to the proton that is cis to the carbonyl oxygen at the time of the acquisition and the higher field hump to the proton trans to the carbonyl oxygen. No other signals in the spectrum of a primary amide will be broadened by restricted rotation about the primary amide bond.

Secondary amides, on the other hand, generally do not exhibit two rotametric forms. That is not to say that rotation about the amide bond in secondary amides doesn't occur at all – just that secondary amides spend most of their time with the two large groups, R, and R^2, trans to each other:

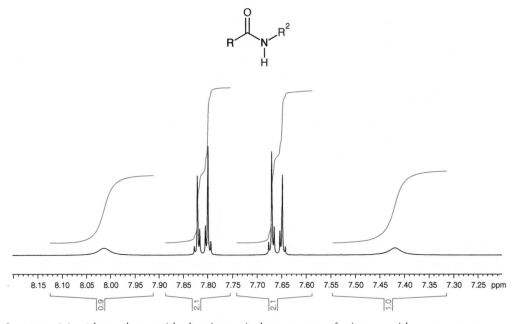

Spectrum 6.8   4-bromobenzamide showing typical appearance of primary amide protons as two non-equivalent broad signals separated by about 0.6 ppm.

For this reason, secondary amides do not generally cause any spectroscopic headaches.

It is the tertiary amides that tend to be the most problematic in terms of proton NMR. They usually exhibit two rotametric forms, the relative proportion of each being determined by both electronic factors and by the relative sizes of the two groups, R^1 and R^2. *Note: This in no way implies that the rotameric forms of a tertiary amide could ever be physically separated as the inter-conversion rate between the two forms is generally in the order of seconds.* A 50/50 ratio of rotamers is only guaranteed where $R^1 = R^2$ (as in the case of a primary amide where $R^1 = R^2 = H$). Consider the following two compounds.

In the case of this molecule, only the protons of the piperidone ring would be affected by restricted rotation about the amide bond. As far as the aromatic protons are concerned, there is no anisotropic difference in the environment they experience, because the piperidone has a plane of symmetry through it.

Now consider:

In this case, there is no such symmetry and so *all* the signals of the spectrum of this compound would be expected to be broadened or duplicated! Always consider the symmetry of the molecule in anticipation of the extent of rotameric complexity.

We will see later on that we can often overcome rotational energy barriers (providing they are not too high) and thus simplify our spectra by running our samples at high temperature. Note that in cases where there is a large difference in the ratio of the rotamers, the coalescence point will not just be midway between the positions of the two rotamers, but will be closer to the position of the major rotamer. Note also that in cases where the amide function is sterically constrained, rotamers may *not* be observed as one rotameric form might be of significantly lower energy than the other and therefore predominate, totally e.g. in a molecule like this:

Another group which is well known for restricted rotation is the nitrovinyl group:

This time, the alkene nominal double bond, has sufficient single bond character to permit a certain amount of rotation, as resonance forms can be drawn e.g.

Another group that frequently – and perhaps surprisingly, in view of secondary amide characteristics – exhibits rotameric behaviour is the secondary carbamate (R–COO–NHR[1]), though the energy barrier to rotation tends to be a little lower than in the amide case.

Finally, it's worth mentioning the formamide group. Although this looks like a special case of a secondary amide, rotamers of different intensity are often seen. Compounds with a formamide attached to an aromatic ring can give particularly complex spectra. Not only does the NH proton couple to the CHO proton, with a coupling of about 2–3 Hz in the 'cis' isomer, and 8–9 Hz in the 'trans' isomer, but, any aromatic protons ortho to the formamide are also split out in the rotamers!

So to sum up, we've seen that restricted rotation can give rise to considerable complexity by broadening or duplication of signals. Indeed, overlap of signals from rotameric pairs is commonplace and can cause further ambiguity. As with any other phenomenon, if it is recognised for what it is, and the spectrum can be interpreted in terms of it, then all well and good. If however, the quality of your spectral information is diminished as a result of it (and remember that you may have more than one site of restricted rotation in a molecule), to the point where you cannot be confident about determining the structure of your compound, then further action must be taken. (Like running your sample hot, or perhaps trying it in D$_4$-methanol for example – this solvent can reduce rotational energy barriers, probably by eliminating intramolecular H-bonding.)

But don't assume that just because your compound exhibits restricted rotation, you must run it hot, to do it justice. Not so! Sometimes, the barrier to rotation is just too high to allow simplification by heating. Remember – it is easier to deal with a spectrum of two, sharp rotamers than a broad semi-coalesced mess.

It is worth noting that while we have restricted discussion in this section to conformational inter-conversion based on the slow rotation of bonds, the concept of 'The NMR Timescale' is equally applicable to other types of inter-conversion such as can sometimes be seen in cyclic systems which may exist in two different conformational forms. For more information about dynamic effects in NMR, see Chapter 12.

6.6 Heteronuclear Coupling

So far, we've considered spin-coupling in considerable detail, but only proton–proton coupling. There are, in fact, over 60 elements having nuclei of one or more of their naturally occurring isotopes which have magnetic moments. This means that they not only have their own NMR spectra – e.g. ^{19}F, ^{31}P, which can be recorded with a suitable spectrometer – but also the capability of coupling with protons. The most notable and obvious feature of heteronuclear coupling, is that no reciprocal coupling is observed in the proton spectrum – because it exists in the spectrum of the heteroatom, of course. In this section, we'll have a look at the heteroatoms of importance, which you are quite likely to encounter, and one or two others, which are less commonly encountered. It might be tempting to think that if your compound contains a heteroatom there should be an imperative to acquire a spectrum for that specific nucleus – but this is not so. The proton spectrum often contains all the confirmation of the heteroatom that you need, as the size and nature of the couplings observed can be quite specific.

We will deal with the spectroscopy of a few of these nuclei in later sections but for now, we will restrict ourselves to the consequences of hetero atoms seen in proton spectroscopy.

6.6.1   Coupling between Protons and ^{13}C

Consider Spectrum 6.9 which shows a $CHCl_3$ singlet plotted at very high intensity.

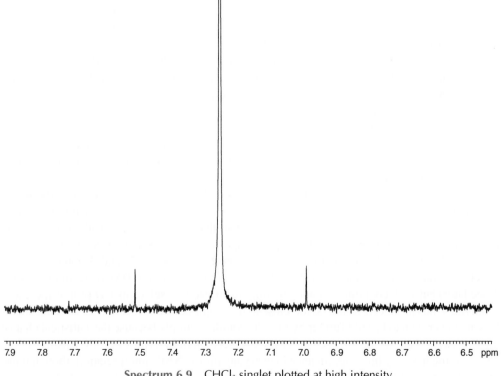

Spectrum 6.9 $CHCl_3$ singlet plotted at high intensity.

On each side of the signal, a number of minor peaks may be seen, one pair of which are the '^{13}C satellites'. (We'll discuss spinning side-bands a little later.) Since the ^{13}C nucleus has a magnetic moment, it couples to proton signals, but as its natural abundance is only 1.1%, the ^{13}C satellites are very small, each satellite accounting for only 0.55% of the intensity of the peak to which it belongs. The only time you might notice them, is when you have a very strong singlet in your spectrum, such as a tertiary butyl.

The ^{13}C nucleus, like the proton, has a nuclear spin quantum number (I) of 1/2, so there are only two permitted energy states of the nucleus with respect to the external magnetic field. This means, of course, that there are only two satellite peaks i.e. the 1.1% of the protons that are attached to ^{13}C nuclei are split by the ^{13}C nucleus into a doublet (and the 98.9% that are attached to ^{12}C, are not). If you measure the coupling (from satellite to satellite), you'll find that it's 210 Hz – though the size of ^{13}C–H couplings vary considerably, depending on the type of function the carbon is incorporated into. This coupling may seem very large, but don't forget it is a 1-bond coupling.

These days, improvements in magnet design and consequent greater field homogeneity have made it quite common practice to run NMR experiments, non-spinning. Indeed, many of the 2-D experiments should definitely *not* be run spinning! However, for 1-D spectra, the best resolution is likely to be obtained by spinning your samples at about 20 Hz. If you do this, you *may* encounter spinning side bands. These should never be a problem in a well-shimmed instrument operating to record spectra at typical levels of gain but it is possible to observe them occasionally as small peaks on either side of very strong peaks (most notable singlets) such as *t*-butyl singlets. Their relative intensities are not fixed as with ^{13}C satellites but can vary with the state of the high-order shims and with the quality of the NMR tubes you use. *Note: Should spinning side-bands ever exceed the size of the ^{13}C satellites, you should seriously consider a major shim of your instrument!* Should you be looking for some very minor constituent of your sample, ^{13}C satellites, and spinning side bands may get in the way. Spinning side bands can be moved by altering the spin rate of the sample tube but you can't do anything about the satellites. Notice that the separation of the first spinning side band (if seen) from the main peak, when measured in Hz, gives the spinning speed (also in Hz of course). Notice too, that the phase of a second spinning sideband, if present, is always 'out' with respect to all the other peaks – a useful diagnostic feature.

^{13}C satellites can actually be quite useful sometimes, as they give a ready-made visual comparator for the quality of spectrometer high-order shimming and for trace impurities that you may be trying to quantify, since we know that each satellite will have an intensity of 0.55% of the peak it is associated with.

Two final interesting points relating to ^{13}C satellites… While they are generally, evenly spaced on either side of the major peak, they do not have to be *exactly* symmetrically disposed about it. It is quite possible to observe a small isotopic shift so that the proton chemical shift of the ^{13}C species is fractionally different from the major ^{12}C species. Also, if you do observe ^{13}C satellites, they will only ever be the product of 1-bond ^{13}C–proton coupling. 2- and 3-bond couplings between ^{13}C and protons certainly exist (and indeed are pivotal in the HMBC technique as we will see later) but such couplings do not generally manifest themselves in 1-D proton spectra as any satellites thus produced would be too close to the major peak to observe. ^{13}C satellites themselves are never seen to be split further by ^{13}C–^{13}C coupling simply because the statistical chance of finding two ^{13}C atoms next to each other is extremely small in terms of NMR sensitivity.

^{13}C coupling has very little significance in everyday proton NMR interpretation, though it has been used in the past to crack specific problems by means of selective enrichment of a specific carbon during synthesis, with a greater than normal percentage of ^{13}C isotope, which makes detection easy.

6.6.2 Coupling between Protons and ¹⁹F

Fluorine usually makes its presence felt in a fairly spectacular fashion, when it is present in a molecule. Once again, $I = 1/2$, so we only have two allowed states to worry about. Unlike ^{13}C, however, fluorine has only one isotope, ^{19}F, and as this of course, has 100% natural abundance, we see the whole proton signal split, instead of a couple of tiny satellites on either side of our signals!

This point is well illustrated with a spectrum of 3-fluoro propanol (Spectrum 6.10), which shows a fairly dramatic example of fluorine coupling. The F–CH₂– coupling is about 47 Hz, and the F–CH₂–CH₂– coupling is 27 Hz. The coupling to the third methylene group is non-existent in this example but can be seen sometimes (0–3 Hz).

Another example of ^{19}F coupling, this time in an aromatic system (4-fluoro benzoic acid), is shown in Spectrum 6.11. Note how the ^{19}F couplings to the aromatic protons give the AA'BB' system an asymmetric appearance. The actual values in this case are 9 Hz (ortho-) and 5.6 Hz (meta-) which are fairly typical.

More useful ^{19}F coupling data is given in Table 6.1.

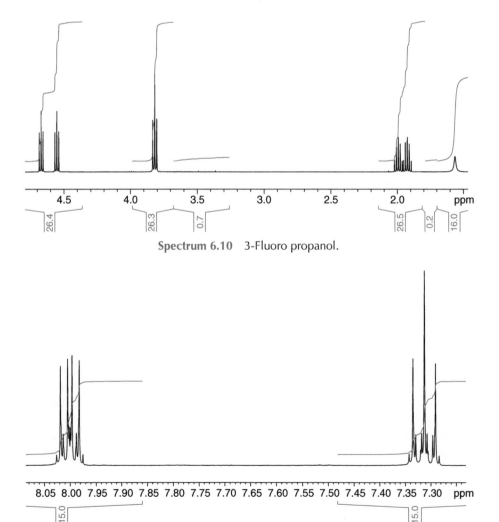

Spectrum 6.10   3-Fluoro propanol.

Spectrum 6.11   4-Fluoro benzoic acid.

Table 6.1   Some typical $^{19}F-$ proton couplings.

Structure	$^{19}F-^{1}H$ position	Typical $^{19}F-^{1}H$ coupling (Hz)
	F–**CH₂**– F–CH₂–**CH₂**– F–CH₂–CH₂–**CH₂**–	45 24 0–3
	F–H (geminal) F–H (cis) F–H (trans)	85 20 50
	F–CH₃	2–4
H~C=C~CF₃	F₃C–H	0–1
	F₃C–CH₂–	8–10
	F–H (ortho) F–H (meta) F–H (para)	6.2–10.3 3.7–8.3 0–2.5
	F–CH₃ (ortho) F–CH₃ (meta) F–CH₃ (para)	2.5 0 1.5
	F$_{axial}$–H$_{axial}$ F$_{axial}$–H$_{equatorial}$ F$_{equatorial}$–H$_{equatorial}$	34 11.5 5–8

Fluorine can sometimes throw up some unexpected couplings in certain situations and spectra need to be handled with care! Sometimes, fluorine can be seen to couple over an unfeasible number of bonds (we have seen a 7-bond coupling in the past!). This is because fluorine is so electron hungry that it can couple through space as well as through bond.

We have also noted some strange behaviour with fluorinated pyridines, for example, 3-fluoro nicotinic acid. See Spectrum 6.12.

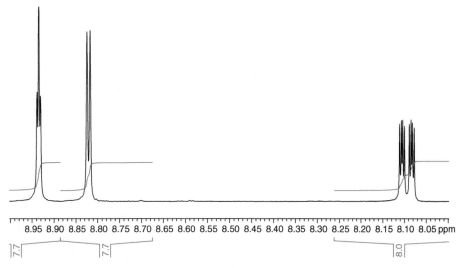

Spectrum 6.12 3-Fluoro nicotinic acid.

The signal for H_c (approx. 8.1 ppm) clearly shows couplings of 9.1, 2.9 and 1.7 Hz. The 9.1 Hz coupling must be from the fluorine as it does not appear anywhere else in the spectrum and its chemical shift distinguishes it from either of the other two protons.

Of the other two protons, the signal at 8.82 ppm (H_b) shows only a 2.9 Hz coupling which is also found in H_c, while H_a exhibits two small couplings (2 and 1.7 Hz), the smallest of these also appearing in H_c. These observations lead to the conclusion that the fluorine–proton couplings in this molecule are as follows:

F–H_a	2 Hz
F–H_b	very small, <1 Hz!
F–H_c	9.1 Hz

F–H_c coupling did not surprise and neither did F–H_a coupling. But the F–H_b coupling of less than a single Hz is totally baffling and defies obvious logic!

Having learnt the lessons from this simple little compound, it would seem reasonable to expect similarly surprising couplings in other fluorinated heterocycles...

Tread carefully!

6.6.3 Coupling between Protons and ^{31}P

Phosphorus is the other heteroatom of major coupling importance to the organic chemist. Like ^{19}F, ^{31}P has a spin of ½ and a 100% natural abundance, so you know what to expect! The actual size of the couplings observed with ^{31}P can vary considerably, depending on the oxidation state of the ^{31}P atom. You'll find some useful examples in Table 6.2.

^{31}P shows one particularly interesting feature. The size of couplings normally decreases dramatically with the number of intervening bonds, but this is not always the case with ^{31}P (Table 6.2).

A proton directly bonded to a ^{31}P atom can be split by an enormous coupling of as much as 700 Hz (depending on the oxidation state of the phosphorus). That means that the two parts of such a signal would be separated by almost 3 ppm at 250 MHz! So huge is this coupling that you could easily fail to recognise or accept it as a coupling at all, if you came across it. Spectrum 6.13 shows an example of ^{31}P–^{1}H coupling.

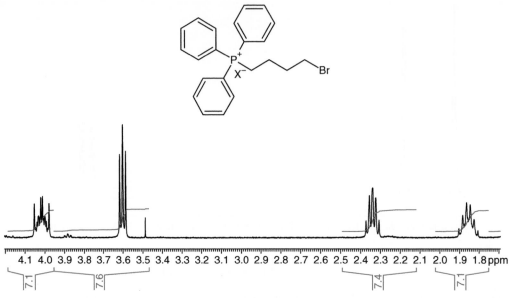

Spectrum 6.13 $^{31}P–^{1}H$ coupling.

Table 6.2   Some typical ^{31}P–proton couplings.

Structure	$^{31}P–^{1}H$ relative position	Typical $^{31}P–^{1}H$ coupling (Hz)
$(CH_3)_3P$	P–CH₃	2.7
$(CH_3)_3P=O$	P–CH₃	13.4
$(CH_3)_4P^+I^-$	P–CH₃	14.4
$(CH_3–CH_2)_3P$	P–**CH₂**–CH₃	0.5!
	P–CH₂–**CH₃**	13.7!
$(CH_3–$	P–CH₂–	11.9
$CH_2)_3P=O$	P–CH₃	16.3

Structure	$^{31}P–^{1}H$ relative position	Typical $^{31}P–^{1}H$ coupling (Hz)
R—P with two H (P–H)	P–H	180–200
RO—P(=O)(OR)—H	P–H	630–710
H₃C—C(R)(R)—P(OR)(OR)	P–CR₂–**CH₃**	10.5–18
H₃C—C(R)(R)—P(OR)(R)	P–CR₂–**CH₃**	10.5–18

Table 6.2 (Continued)

Structure	^{31}P–^{1}H relative position	Typical ^{31}P–^{1}H coupling (Hz)
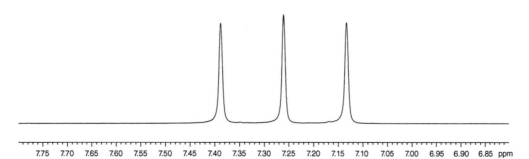	P–CR$_2$–**CH$_3$**	10.5–18
	P–**CH$_2$**–	6

The complex multiplet at 4.02 ppm shows a 13 Hz 2-bond ^{31}P coupling to the first –CH$_2$– in the chain and spin decoupling enables the 3-bond ^{31}P coupling to the next –CH$_2$– in the chain (1.86 ppm) to be measured (8 Hz).

6.6.4   Coupling between ^{1}H and Other Heteroatoms

If you ever run a sample, which is contaminated with an ammonium salt, in DMSO, you will see ^{14}N–proton coupling, as shown in Spectrum 6.14. Note that the three lines of the multiplet are of equal intensity (the middle line is a little bit taller than the outer ones, but this is because of the width of the peaks at their bases. The central signal is reinforced because it stands on the tails of the outer two). This is because ^{14}N has a spin of I=1, and the allowed states are therefore -1, 0 and $+1$. This three-line pattern with its 51 Hz splitting is highly characteristic and once seen, should never be forgotten.

^{14}N coupling is only observed when the nitrogen atom is quaternary. In all other cases, any coupling is lost by exchange broadening, or quadrupolar broadening, both of which we've discussed before. 2-bond couplings (e.g. $^{14}N^{+}$–(**CH$_2$**)$_4$) are not observed, even when the nitrogen is quaternary, in 'quat-salts' such as (n-butyl)$_4N^{+}$ Br–, presumably because the coupling is very small. So the phenomenon is only ever observed in the $^{+}NH_4$ ion! *Note: The –CH$_2$– attached to the quaternary nitrogen in compounds like tetra N-butyl ammonium chloride **does** present as a distorted triplet with its central line split into a narrow triplet but this has nothing to do with ^{14}N coupling as the same distortion can sometimes be seen in –CH$_2$– groups next to certain other moieties, e.g. –SO$_2$R. It is a non-first-order phenomenon caused by slight non-equivalence of the two protons in question.*

7.75 7.70 7.65 7.60 7.55 7.50 7.45 7.40 7.35 7.30 7.25 7.20 7.15 7.10 7.05 7.00 6.95 6.90 6.85 ppm

Spectrum 6.14   Typical appearance of $^{+}NH4$ ion in DMSO.

Boron has two isotopes, both of which have spin! ^{10}B has a natural abundance of 18.8%, and a spin of $I = 3$ (allowed spin states $-3, -2, -1, 0, +1, +2, +3$ i.e. one signal will be split into seven lines of equal intensity), while ^{11}B has a natural abundance of 81.2%, and a spin of $I = 3/2$ (allowed spin states $-3/2, -1/2, +1/2$ and $+3/2$. i.e. one signal will be split into four lines of equal intensity).

This gives rise to amazing effects in the borohydride, BH_4^- ion (Spectrum 6.15), which can sometimes be formed accidentally during borohydride reductions. Note that the ^{10}B–H couplings are of a different size to the ^{11}B–H couplings. All 11 lines of the BH_4^- ion are to be found between 0 and -0.7 ppm. Note that like ^{14}N, ^{11}B has a quadrupolar nucleus, but once again, the symmetrical environment of the borohydride ion negates the relaxation pathway that would otherwise cause significant line broadening. Boron coupling is not generally seen in asymmetric environments or over multiple bonds.

One other heteroatom worth mentioning is tin as organo-tin compounds are significant in organic synthesis. Tin has no fewer than ten naturally occurring isotopes, but fortunately, only three of them have nuclear spin. ^{115}Sn has a natural abundance of a mere 0.32%, which makes it spectroscopically insignificant, of course. The only isotopes of tin that need concern us, are ^{117}Sn (natural abundance 7.67% and $I = 1/2$), and ^{119}Sn (natural abundance 8.68%, and also $I = 1/2$).

These two isotopes are both capable of 2-bond and 3-bond couplings in alkyl organo-tin compounds. This is demonstrated in Spectrum 6.16 which shows a mixture of two organo-tin compounds. The compound with a strong central peak at 0.5 ppm is thought to be $(CH_3)_3$–Sn–

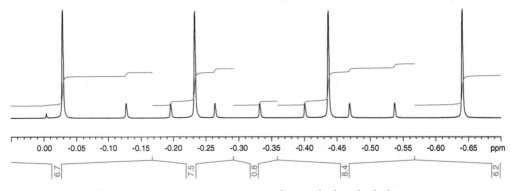

Spectrum 6.15   Boron–proton coupling in the borohydride ion.

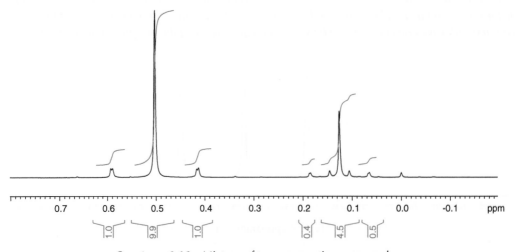

Spectrum 6.16   Mixture of two organo-tin compounds.

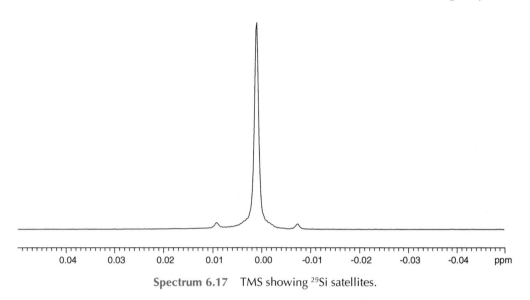

Spectrum 6.17 TMS showing ^{29}Si satellites.

OH. The inner satellites result from a ^{117}Sn–CH$_3$ coupling of 69 Hz and the outer satellites to a ^{119}Sn–CH$_3$ coupling of 72 Hz.

The second compound with the major signal centred at 0.13 ppm is (CH$_3$)$_3$–Sn–Sn–(CH$_3$)$_3$. In this case, we see once again, satellites resulting from 2-bond couplings but also a second set of inner satellites resulting from smaller 3-bond couplings of about 16 Hz for both ^{117}Sn and ^{119}Sn (i.e. **Sn–Sn–CH$_3$**).

Note too from the chemical shifts of these methyl groups that tin has quite a strong shielding effect.

Finally, ^{29}Si is an isotope that you should be aware of – every time you acquire a well prepared sample using TMS as a standard! ^{29}Si satellites (accounting for about 4.7% of the total signal, J ^{29}Si–CH$_3$, 6.6 Hz) should be visible at the base of your TMS peak. The small coupling provides a good test of shimming quality (Spectrum 6.17).

6.7 Cyclic Compounds and the Karplus Curve

As we have already mentioned, chemical shifts and couplings are heavily influenced by molecular constraint and for this reason, some guidance in dealing with cyclic (saturated) compounds might well prove useful. We have already seen that in straightforward open-chain alkyl systems, the size of proton–proton couplings is governed by the electronegativity of neighbouring atoms. But the most important factor which governs the size of couplings between vicinal protons is the dihedral angle between them.

In open-chain systems, this angle is usually averaged by rotation about the C–C single bond, and so is not normally of significance. But in carbocyclic systems, dihedral angles are usually fixed, since the structures are generally rigid. It is therefore vital that we understand how the size of vicinal couplings varies with dihedral angle. This data can be obtained by using the Karplus equation but the information derived from this equation (or equations as there are various versions of it) is more usefully portrayed graphically. A family of curves thus constructed makes additional allowance for factors other than dihedral angles which influence vicinal proton couplings e.g. localised electronegativities (Figure 6.3) but we have opted for a simplified graph showing only three curves.

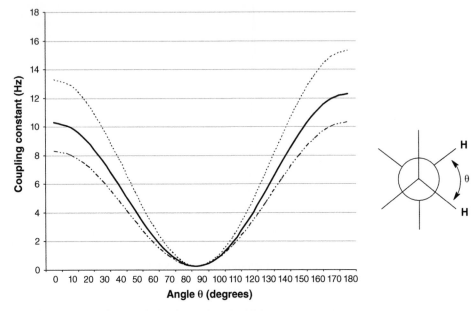

Figure 6.3   The Karplus curve.

Selection of the best curve for a given situation is perhaps rather a matter of trial and error, but is best approached by positively identifying an axial–axial coupling, since this arrangement ensures (in six-membered rings at least) a dihedral angle of 180° between the protons. Choose the curve that best fits the value that you observe for an axial–axial coupling in your molecule. Note that in the absence of any extreme electronic effects, this should give rise to a coupling of about 12 Hz. Similarly, a dihedral angle of 90° gives rise to a coupling of approx. 0 Hz, and where the angle is 0°, we may expect a coupling of about 10 Hz. Making a model of the molecule becomes very important in the case of carbocyclic compounds, as it is important to be able to make fairly accurate estimates of dihedral angles.

Now let us consider Spectrum 6.18 and the structure that gave rise to it and see how the 'Karplus curve' can be used to aid assignment of the spectrum. (This compound will be referred to from now on as *the* morpholine compound as we will use it to demonstrate several different techniques.) Note that the aromatic region has been omitted as it contains little of interest and we wish to concentrate on the carbocyclic region of the spectrum. It was acquired in $CDCl_3$.

To derive maximum benefit from this exercise, we recommend that you make a model of this molecule, and refer to it as we go through the spectrum. Note that the morpholine ring falls naturally into a 'chair' conformation. Note also that in this example, the $-CH_2-Cl$ function will seek to minimise the morpholine ring energy by occupying an equatorial environment as this mini-

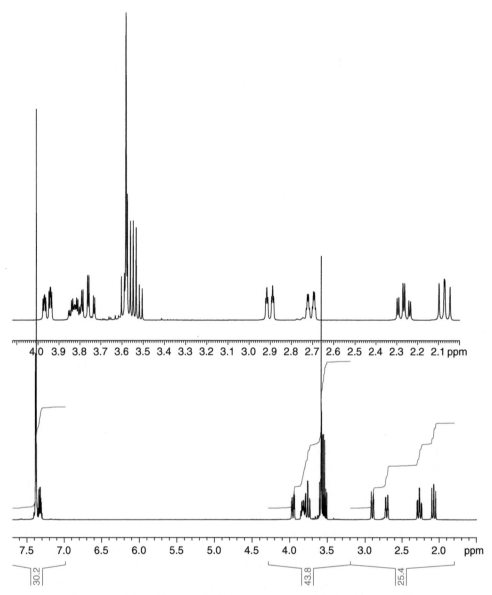

Spectrum 6.18 The morpholine compound in CDCl$_3$ with expansion.

mises steric interactions between it and protons, and other substituents on the ring. All groups do this. The benzyl function will do likewise by inversion of the nitrogen stereochemistry.

It is also worth noting that nine times out of ten, equatorial protons are observed at somewhat lower field than the corresponding axial protons. This can be reversed in certain cases where the specific anisotropies of the substituents predominate over the anisotropies of the rings themselves but this is relatively rare. The difference is typically 0.5–1 ppm, but may be more.

The structure is depicted as a Newman projection (Figure 6.4). Aromatic protons aside (they give the expected five-proton multiplet centred at around 7.3–7.4 ppm), the first signal we encounter as we work from left to right is a complex multiplet – which is actually, a doublet of doublet of doublets [ddd] – at 3.95 ppm. Careful measurement of the couplings reveals them to

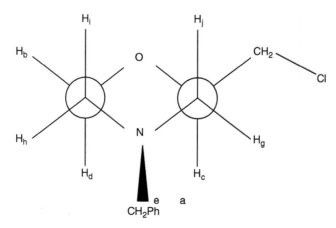

Figure 6.4   The morpholine compound shown as a Newman projection.

be 11.4, 3.4 and 2.0 Hz. Since the multiplet is dominated by one large coupling, we can be safe in the knowledge that it must be an *equatorial* proton.

This is because the dihedral angles between equatorial protons and both their equatorial and axial vicinal partners are always such that they give rise to relatively small couplings (check model and the Karplus curves). The only large coupling (i.e. 10 Hz or more) an equatorial proton can have will always be to its geminal partner – if it has one. So in this case, the 11.4 Hz coupling is clearly a geminal coupling. If we now make the entirely reasonable deductions that the proton giving rise to this signal is likely to be alpha to oxygen rather than nitrogen (on the basis of chemical shift) and that as the –CH$_2$–Cl will be equatorial (as explained earlier), then this multiplet can only be assigned to the equatorial proton 'b' since there are no other equatorial protons that are alpha to oxygen in the molecule. The other two couplings can be rationalised in terms of the equatorial–axial coupling (3.4 Hz which is reciprocated in the ddd at 2.27 ppm) and the equatorial–equatorial coupling (2 Hz which is reciprocated in the dddd at 2.71 ppm). *Note that methods of un-picking couplings will be discussed at length in later sections. Such methods are very useful when dealing with more complex spin systems like this one.*

The degree of roofing of 'b' indicates that its geminal partner must be fairly close to it in terms of chemical shift and sure enough, the six-line multiplet (another ddd) centred at 3.76 ppm. satisfies the requirements for this proton ('d'). Note that the second-large coupling to this signal is also 11.4 Hz, the axial–axial coupling being the same size as the geminal coupling in this instance. The small remaining coupling (approx. 2 Hz) is reciprocated in the dddd at 2.71 ppm and is an axial–equatorial coupling.

Proton 'c' can be defined by the fact that it is *not* equatorial and it is highly coupled. The multiplet at 3.82 ppm satisfies these requirements. It is in the right ball park for chemical shift and is highly complex in that this proton is already the X-part of an ABX system coupled to both protons alpha to the chlorine (the AB part). It is then further coupled with a 10 Hz, axial–axial coupling (reciprocated in the dd at 2.07 ppm) and with a 2 Hz axial–equatorial coupling which is reciprocated in the ddd at 2.90 ppm. Note that 'c' and 'd' are not fully resolved from each other. Such overlap inevitably complicates the issue.

The *N*-benzyl protons are accidentally equivalent, presenting as a singlet at 3.59 ppm and overlap with the two protons alpha to the chlorine atom which present as the heavily roofed AB part of an ABX system (i.e. eight lines) centred at 3.55 ppm.

Without slavishly dissecting the remaining four signals (2.90, 2.71, 2.27 and 2.07 ppm), we hope that the principles of carbocyclic analysis have now been established. You should see at a glance that the 2.90 and 2.71 ppm signals must belong to equatorial protons because they are each dominated by only one large coupling and the remaining two must correspond to their axial partners. You should now be able to verify which equatorial proton belongs to which axial proton just by inspection.

There is one last coupling which we have not yet mentioned and that is the apparent extra small coupling that can be seen on the equatorial protons alpha to the nitrogen (2.90 and 2.71 ppm). These two signals are in fact coupled to each other by what is known as a W-path coupling. These are 4-bond couplings (unusual in saturated systems) which can be seen in situations where all the intervening proton–carbon and carbon–carbon bonds lie in the same plane. You can see from the model which you have next to you (?) that by definition, such protons can only be equatorial. Note that while all the assignments in this section have been made purely on the basis of observations of couplings and multiplet appearance, this type of assignment is often simplified by having definitive knowledge of coupling pathways. We will discuss the options available for acquiring this type of data in a later chapter.

While six-membered rings may often give rise to quite complex spectra, they are at least generally rigid and based on the 'chair' conformation. As we have seen, this means that dihedral angles can be relied on and the Karplus curve used with reasonable confidence. Unfortunately, however, the same approach will end in tears if applied to other ring systems. Five-membered rings, for example, are notoriously difficult to deal with as they have no automatic conformational preference. They are inherently flexible, their conformations driven by steric factors. Cis protons on five-membered rings can have dihedral angles ranging from approximately −30 to 0 to +30 degrees an exhibit a range of couplings to match. Trans protons on the other hand can range from +90 to +150 degrees. Deductions that can be made on the basis of observed vicinal couplings are therefore limited. If the observed coupling is *very* small, the two protons can only be trans to each other but if it is not, then they may be either cis or trans. We counsel against reliance on molecular modelling packages to produce a valid conformation of such structures. The energy difference between potential conformers is often small and could change in different solvents. Modelling packages consider molecules in isolation and thus make no allowances for solvent effects. Stereochemical assignments of such ring systems can only be confidently made on the basis of NOE experiments in combination with an HSQC experiment. We will be looking at both these very important techniques in detail later on but briefly, this preferred approach entails the positive identification of all the protons on the ring and in particular, the linking of geminal pairs by HSQC. Ring protons and proton-bearing groups on each face of the ring can then be inter-related by NOE experiments and a picture of the ring built up.

6.8 Salts, Free Bases and Zwitterions

Sometimes, misunderstandings can arise when dealing with compounds containing protonatable centres. Hopefully, in this section, we will be able to clarify a few key issues that are relevant to such compounds.

As we have already mentioned, $CDCl_3$ should be avoided as a solvent for salts for two reasons. Firstly, salts are unlikely to be particularly soluble in this relatively non-polar solvent but more importantly, spectral line shape is likely to be poor on account of relatively slow proton exchange at the protonatable centre. The remedy is simple enough – avoid using $CDCl_3$ and opt for one of the more polar options instead e.g. deuterated DMSO or MeOH and you should obtain spectra every bit as sharp as those of free bases.

In practical terms, it is invariably a nitrogen atom that is protonated in salt formation. This always leads to a downfield shift for protons on carbons both alpha and beta to the nitrogen concerned. In alkyl amines, the expected shifts would be about 0.7 and 0.3 ppm respectively. Remember that some heterocyclic compounds (e.g. pyridine) contain nitrogen atoms that are basic enough to protonate and comparable downfield shifts can be expected (Spectrum 5.9).

A misconception that we commonly encounter is that a spectrum can be a 'mixture of the salt and the free base'. This is an excuse that is often used by chemists to explain an inconveniently messy-looking spectrum. Don't be tempted by this idea – proton transfer is fast on the NMR timescale (or at least, it is when you use a polar solvent) and because of this, if you have a sample of a compound that contains only half a mole-equivalent of an acid, you will observe chemical shifts which reflect partial protonation and NOT two sets of signals for protonated and free-base forms. It doesn't happen – ever!

Of course, whether a compound forms a salt or not depends on the degree of availability of the lone pair of electrons on any nitrogen atoms in the compound (i.e. their pKbs) and on how strong the acids involved in the salt formation (pKas). As a rough rule-of-thumb, alkyl and aryl amines *do* form salts while amides, ureas, most nitrogen-containing heterocycles and compounds containing quaternary nitrogen atoms *do not*.

It should always be remembered, of course, that the NMR spectrum reflects a compound's behaviour *in solution*. It is quite possible for a compound and a weak acid to crystallise out as a stoichiometric salt and yet, in solution, for the compound to give the appearance of a free base. For this reason, care should be taken in attempting to use NMR as a guide to the extent of protonation. If the acid has other protons that can be integrated reliably e.g. the alkene protons in fumaric or maleic acid, then there should be no problem but if this is not the case e.g. oxalic acid, then we would counsel caution! Do not be tempted to give an estimate of acid content based on chemical shift. With weak acids, protonation may not occur in a *pro rata* fashion though it is likely to in the case of strong acids.

Sometimes, you may encounter compounds which have more than one protonatable centre. It is often possible to work out if either one or more than one are protonated in solution. A good working knowledge of pKbs is useful to help estimate the likely order of protonation with increasing acidity. Assume that the most basic centre will protonate first and assess the chemical shifts of the protons alpha to each of the potentially protonatable nitrogen atoms.

6.9   Zwitterionic Compounds Are Worthy of Special Mention

$$H_2N - R - COOH \leftrightarrow H_3N^+ - R - COO^-$$

By their very nature, their partial charge separation can make them fairly insoluble and the degree of this charge separation (and hence resultant NMR spectra) tends to be highly dependent on concentration and pH. For these reasons, we recommend dealing with such compounds by 'pushing' them one way or the other i.e. by adjusting the pH of your NMR solution so that the compound in question is either fully protonated (addition of a drop of DCl) or de-protonated (addition of a drop of saturated sodium carbonate in D_2O).

$$H_3N^+ - R - COOH \qquad H_2N - R - COO^-$$
Fully protonated   Fully de-protonated

While dealing with protonation issues, it is well worth considering the time-dependence of the process in the context of the NMR timescale. A compound of the type shown below provides an interesting example.

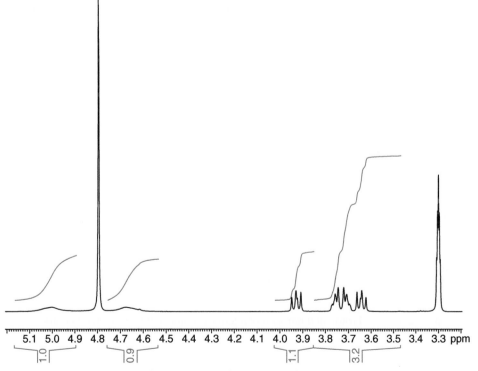

As a free base, the Ar–CH$_2$–N protons would present themselves as a simple singlet. The lone pair of electrons on the nitrogen invert very rapidly on the NMR timescale and so the environment of the two protons is averaged and is therefore identical. However, on forming a salt, the whole process of stereochemical inversion at the nitrogen is slowed down dramatically because the sequence of events would be de-protonation, inversion and re-protonation. Although as we said earlier, proton transfer is in itself a very fast process on the NMR timescale, it is the time taken for the *entire* process to occur that determines the nature of the spectrum that we observe.

What we actually observe for the Ar–CH$_2$–N protons of the salt in this molecule is a pair of broad, featureless signals at 4.6 and 5.0 ppm. The explanation for this is simple enough once the concept of time dependency for the inversion sequence has been appreciated. The protons in question find themselves in different environments (within the context of the NMR timescale) and therefore have distinct chemical shifts. The signals are broad because the dynamic exchange process is taking place with a time period comparable to the NMR timescale, the broadening masking the geminal coupling between them. See Spectrum 6.19.

Spectrum 6.19   Slow inversion of a protonated tertiary amine nitrogen.

A logical extension of these ideas will lead you to a recognition of the fact that a phenomenon of this type could yield species in solution which appear to behave as if they contain a chiral centre – even when they don't. We have seen 'pseudo enantiomeric' behaviour in compounds of the type shown below (when protonated).

$$A = Aryl$$

All the protons of the CH_2s in a molecule of this type may be non-equivalent (i.e. you observe essentially three AB systems). Note that the coupling from the alkene CH is would be small to both of the cyclic CH_2s when the spectrum is acquired in the presence of HCl. See Spectrum 6.20. When the free base is liberated, all the AB systems collapse to give singlets. The explanation follows on logically from a consideration of the previous example. Protonation of the tertiary amine generates a chiral centre at the nitrogen atom, forcing all the geminal pairs of protons into different environments – hence the three AB systems. But this does not in any way imply that it would be possible to separate out enantiomers of the compound in salt form. These 'pseudo enantiomers' can only be differentiated within the context of a technique which has a timescale of a couple of seconds. Attempting to separate them on an HPLC column, for example, would be unsuccessful as this technique has a timescale of several minutes (defined by how long compounds take to travel down the column and enter the detector). During this time, proton exchange and consequent 'enantiomer' inter-conversion would have occurred many times in the course of the analysis. The only manifestation of this might be a slightly broader than normal (single) peak.

This whole area of spectroscopy touches on many different topics and can only be approached confidently with a reasonable working knowledge of basic NMR, stereochemistry and certain aspects of physical chemistry.

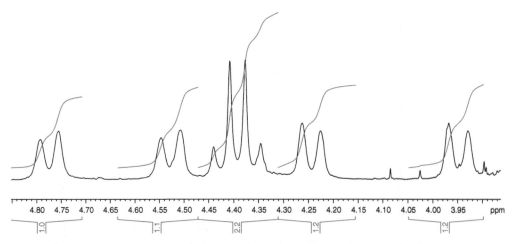

Spectrum 6.20   Protonated nitrogen of a tertiary amine acting as a 'chiral centre'.

7

Further Elucidation Techniques – Part 1

If a spectrum does not yield the definitive information that you require on inspection, there are many other 'tools-of the-trade' that we can use to further elucidate structures. Broadly speaking, these fall into two categories – chemical techniques and instrumental techniques.

We will take a brief look at chemical techniques first. It is true to say that the development of more and more sophisticated instrumental techniques has, to a considerable extent, rendered these less important in recent years but they still have their place and are worthy of consideration in certain circumstances.

Before embarking on any of these chemical techniques, however, be advised that they are not without a measure of risk as far as your sample is concerned. One of the great strengths of NMR is that it is a non-destructive technique but that can change quite rapidly if you start subjecting your compounds to large changes in pH or to potentially aggressive reagents like trifluoroacetic anhydride! Be sure that you can afford to sacrifice the sample as recovery may not be possible. In the case of precious samples, chemical techniques should be regarded as a last option rather than a first choice.

7.1   Chemical Techniques

7.1.1   Deuteration

Deuteration is the most elementary of the chemical techniques available to us, but it is still useful for assigning exchangeable protons which are not obviously exchangeable, and for locating exchangeables masked by other signals in the spectrum. There are of course, other ways of identifying exchangeables. The signal can be scrutinised closely to see if it has any ^{13}C satellites associated with it, though this is not viable in the case of broad signals. Alternatively, irradiation of the water peak in an NOE experiment can be used as we'll see later. Nonetheless, deuteration does provide a quick and easy method of identification which is still perfectly valid.

Just to recap on the procedure, add a couple of drops of D_2O to your solution, and shake vigorously for a few seconds. Note that with $CDCl_3$ solutions, the best results are obtained by passing the resultant solution through an anhydrous sodium sulfate filter to remove as much emulsified D_2O as possible. (Note also that $CDCl_3$ and D_2O are not miscible, the $CDCl_3$ forming

---

Essential Practical NMR for Organic Chemistry, Second Edition. S.A. Richards and J.C. Hollerton.
© 2023 John Wiley & Sons Ltd. Published 2023 by John Wiley & Sons Ltd.

the bottom layer as it is denser than D_2O.) You then re-run the spectrum and check for the disappearance of any signals. Careful comparison of integrations before and after addition of D_2O should reveal the presence of any exchangeable protons buried beneath other signals in the spectrum. If they are slow to exchange, like amides, a solution of sodium carbonate in D_2O, or NaOD may be used. The technique is demonstrated using *n*-butanol in Spectrum 7.1 and Spectrum 7.2. Note the reduction in integration of the multiplet centred at 1.39 ppm.

Remember, as was mentioned previously, that any proton which is acidic enough is prone to undergo deuterium exchange. Methylene protons alpha to a carbonyl, for example, may

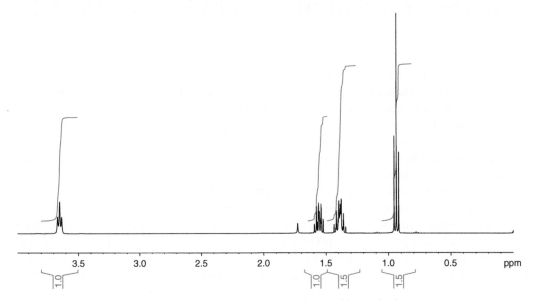

Spectrum 7.1   *n*-Butanol in CDCl₃ with –OH obscured by multiplet at 1.39 ppm.

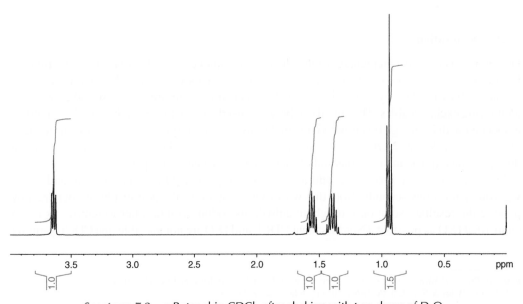

Spectrum 7.2   *n*-Butanol in CDCl₃ after shaking with two drops of D_2O.

exchange if left standing with D_2O for any length of time, as they can exchange via the keto-enol route, i.e.

Note that deuterium exchange of the –OH leads to incorporation of deuterium alpha to the carbonyl in the ketone form. This may happen, even if there is no evidence of any enol signals in the spectrum initially, i.e. it can occur even when the equilibrium is heavily in favour of the ketone. Aromatic protons of rings which bear two or more –OH groups are also prone to undergo slow exchange, as are nitrovinyl protons.

7.1.2   Basification and Acidification

This topic has been dealt with quite extensively in Section 6.9 so we won't go over the material again but there is perhaps one other type of problem that may be worth looking at with a view to solving by a change of pH. Consider the two structures below:

While the preferred method of differentiating these structures would be by an NOE experiment, it would be possible to accomplish this by running them in DMSO and then adding a drop of base to each solution and re-running. (Note: DMSO is the preferred solvent for this experiment as both the neutral and the charged species would be soluble in it.) In both cases, the phenoxide ion (Ar–O⁻) would be formed and the extra electron density generated on the oxygen would feed into the ring and cause a significant upfield shift of about 0.3–0.4 ppm in any protons ortho- or para- to the hydroxyl group. In the example above, the compound on the left would show such an upfield shift for only a single doublet (a), while the compound on the right would show an analogous upfield shift for both a narrow doublet (b) and a doublet of doublets (c).

Caution should be exercised if attempting any determination of this type as it is not the preferred method and it is always safest if *both* compounds to be distinguished are available for study in this way.

7.1.3   Changing Solvents

If a signal of particular interest to you, is obscured by other signals in the spectrum, it is often worth changing solvent – you might be lucky, and find that your signal (or the obscuring signals) moves sufficiently to allow you to observe it clearly. You might equally well be unlucky of course, but it's worth a try.

Running a sample in an anisotropic solvent, like D_6-benzene or D_5-pyridine, can bring about some even more dramatic changes in chemical-shifts. We tend to use benzene in a fairly arbitrary fashion, but in some cases, there is a certain empirical basis for the upfield and downfield shifts we observe.

For example, benzene forms collision-complexes with carbonyl groups, 'sitting' above and below the group, sandwich-style. When the carbonyl is held rigidly within the molecule, either because it forms part of a rigid system, or because of conjugation, we can generally expect protons on the oxygen-side of a line drawn through the carbon of the carbonyl, and at right angles to the carbonyl bond to be deshielded. Conversely, those on the other side of the line are shielded.

7.1.4 Trifluoroacetylation

This is quite a useful technique which can give a rapid, positive identification of –OH, –NH$_2$ and –NHR groups in cases where deuteration would be of little value. Even though the technique can be a little time-consuming and labour-intensive in terms of sample preparation, it can nonetheless yield results in less time than it would take to acquire definitive ^{13}C data – particularly if your material is limited.

Consider Spectrum 7.3. The bottom trace shows the ordinary spectrum of cyclohexanol, run in CDCl$_3$. Distinguishing it from chlorocyclohexane is not easy (without the use

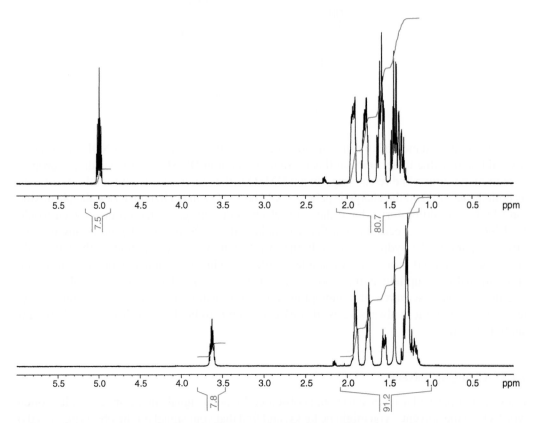

Spectrum 7.3   The use of trifluoroacetic anhydride to identify an –OH group.

of ^{13}C NMR) – the chemical shift of the proton alpha to the functional group would be similar in both compounds, and in the case of the alcohol, the –OH may not show coupling to it. Furthermore, in problems of this type, the –OH proton itself may well be obscured by the rest of the alkyl signals or combined with the solvent water peak. Integration of the alkyl multiplet before and after deuteration will not necessarily be very reliable, since looking for one proton in a multiplet of 10 or 11, will give only a relatively small change in integral intensity (and let us not forget that water in the CDCl$_3$ which will absorb in this region, along with any water that may be residual in the compound).

The top trace shows what happens when the sample is shaken for a few seconds with a few drops of trifluoroacetic anhydride (TFAA). The following reaction occurs:

The resultant spectrum is clearly very different from the alcohol, as the trifluoroacetic ester function is far more deshielding with respect to the alpha proton than is the –OH group. A downfield shift of more than 1 ppm can be seen. This clearly distinguishes the alcohol from the analogous chloro-compound which would of course give no reaction.

This is a relatively quick and convenient technique, the reagent reacting quite readily (assuming no great steric hindrance of course) with alcohols, primary and secondary amines. (Note the possibility of complication if you react a secondary amine with TFAA – it will yield a tertiary amide!) If the reaction is a little slow, as is often the case with phenolic –OH groups, you can 'speed it up a little' by gentle warming, more shaking and even adding a drop of D$_5$-pyridine to base-catalyse the reaction.

Use of this reagent is, however, somewhat limited. You can only use it in solvents which don't react with it (D$_4$-methanol, and D$_2$O are obviously out of the question), or contain a lot of water, i.e. D$_6$-DMSO. Another slight drawback is that the cleaving of the anhydride liberates trifluoroacetic acid, which has nuisance value if your compound is very acid-sensitive, and will also protonate any unreacted amine function present. If this salt-formation is a problem, it is worth adding sodium bicarbonate in D$_2$O, dropwise, to neutralise the acid. You'll know when the excess TFA has been neutralised, as further additions of bicarbonate fail to produce any further effervescence (shake thoroughly, and don't forget to release pressure built-up by release of carbon dioxide!) Dry your solution through an anhydrous sodium sulfate filter before re-running.

7.1.5 Lanthanide Shift Reagents

Unfortunately, the use of lanthanide shift reagents such as the europium compound, Eu(fod)$_3$, is a practice that has been largely consigned to the dustbin of history so we will say very little about them. The problem with trying to use them in high-field spectrometers is that the fast relaxation times of the collision complexes brought about by paramagnetic relaxation (courtesy of the europium or other lanthanide atom), leads to severe line broadening. This paramagnetic broadening is very much worse in high field spectrometers so if you are using a 250 MHz or above, don't bother trying.

If, however, you are still soldiering on with a spectrometer of 100 MHz or less, then by all means try using them to 'stretch' a spectrum out – if your compound is suitable. They work by co-ordinating with an atom that has a lone pair of electrons available for donation. The more available the lone pair, the greater will be the affinity ($-NH_2$/NHR > –OH >> C=O > –O– > – COOR > –CN). Note that they will only work in dry solvents that don't contain available lone pairs. Good luck!

7.1.6   Chiral Resolving Agents

We have seen that the spectra of enantiomers, acquired under normal conditions, are identical. The NMR spectrometer does not differentiate between optically pure samples and racemic ones. The wording is carefully chosen, particularly 'normal conditions', because it is often possible to distinguish enantiomers, by running their spectra in abnormal conditions – in the presence of a chiral resolving agent. Perhaps the best known of these is (–)2,2,2,trifluoro-1-(9-anthryl) ethanol, abbreviated understandably to TFAE (W.H. Pirkle and D.J. Hoover, *Top. Stereochem.*, 1982, 13, 263). Its structure is shown below:

This reagent may form weak collision-complexes with both enantiomers in solution. As the reagent is itself optically pure, these collision-complexes become 'diastereomeric', i.e. if we use (–)TFAE (note that the (+) form can be used equally well), the complexes formed will be (–)reagent – (+)substrate, and (–)reagent – (–)substrate. These complexes often yield spectra sufficiently different to allow both discrimination and quantification of enantiomers. This difference will be engendered largely by the differing orientations of the highly anisotropic anthracene moieties in the two collision complexes. You won't be able to tell which is which by NMR, of course – that's a job for polarimetry or circular dichroism, but if you know which enantiomer is in excess, you can get a ratio, even in crude samples which would certainly give a false reading in an optical rotation determination.

The use of TFAE is demonstrated in Spectrum 7.4, which shows the appearance of proton 'b', before and after the addition of 30 mg of (+)TFAE to the solution. (This is the region of interest – it is usually protons nearest the chiral centre, which show the greatest difference in chemical shifts in the pair of complexes formed.) The middle trace shows the expansion of proton 'b' prior to the addition of TFAE, and the top trace its appearance after the addition of the reagent. It is clear from this trace that proton 'b' is no longer a simple X-part of an ABX system. Apart from shifting upfield, it has broadened, and eight lines are apparent. This is because the 'b' protons in the two collision complexes have slightly different chemical shifts and we can now see them resolved from one another. (Clearly, this sample was a racemate as the ratio of the

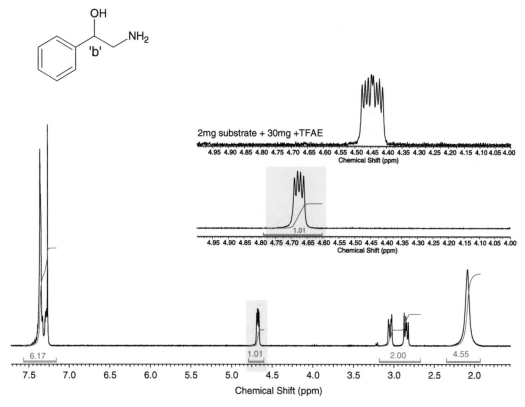

Spectrum 7.4   The use of TFAE as a chiral resolving agent.

resolved 'X' multiplets is 50/50.) This difference is often quite small, and so as to exploit it to the full, experiments of this type are best performed at high field (e.g. 400 MHz or more).

TFAE is a very useful reagent for this type of work, as it is very soluble in CDCl₃, which is just as well, as a considerable quantity of it is often needed to produce useful separations – even at high field. Its other advantage is that its own proton signals are generally well out of the way of the sort of substrate signals you are likely to be looking at. It should be noted that the compound under investigation should have at least one, and preferably two potentially lone-pair donating atoms to facilitate interaction with the reagent.

Another useful reagent of this type is 'chiral binaphthol' (see below).

(R)-(+)-1,1′-Bi-2-naphthol

This is a member of an interesting class of compounds which are chiral, without actually containing a defined chiral centre (known as 'atropisomers'). They are chiral because their mirror images are non-superimposable. In the case of this molecule, there is no rotation about the bond between the two naphthol rings because of the steric interaction between the two hydroxyl groups. 'd' and 'l' forms can be isolated and are perfectly stable (*Analytical Applications of Spectroscopy*, edited by C. S. Creaser and A. M. C. Davies, 1988, p. 346. Optical Purity Determination by 'H NMR, D. P. Reynolds, J. C. Hollerton and S. A. Richards).

Optically pure mandelic acid (see below) can be a useful chiral resolving agent where the compound you are looking at has a basic centre, as it can form an acid-base pair with it, which is a stronger form of association. This compound is of sparing solubility in CDCl$_3$ however, and can precipitate-out your compound, if, as is often the case, its protonated form is of low solubility in CDCl$_3$.

Mandelic acid

The technique of using resolving agents is obviously a useful one in following the synthesis of a compound of specified chirality. To summarise, we take our compound, which has a chiral centre of unknown rotation and form some sort of complex by introducing it to a reagent of known optical purity. The complexes we form have diastereoisomeric character, which can give rise to a difference in the chemical shifts of one or more of the substrate-signals. This enables us to determine the enantiomer ratio, either visually, or by integration, if we have sufficient signal-separation.

In practice, if using one of these reagents to follow the course of a chiral separation, it is essential to determine whether resolution is possible, by performing a test experiment either on a sample of racemate, or at least a sample known to contain significant quantities of both enantiomers. Once useable resolution has been established, the technique can be used to monitor solutions of unknown enantiomer ratios with reasonable accuracy, down to normal NMR detection limits.

It is a good idea to keep the ratio of reagent to sample as high as possible. We recommend starting with about 1–2 mg of compound in solution with about 10 mg of reagent. In this way, you can minimise both the quantity of your sample and the amount of (expensive) reagent used. Keeping the initial sample small has another advantage – it avoids line broadening associated with increased viscosity of very concentrated solutions, while at the same time leaving the option open for further increasing the concentration of reagent, if needed.

One final point – the use of chiral resolving agents is restricted to non-polar solvents, i.e. CDCl$_3$ and C$_6$D$_6$, though combining these can sometimes augment separation.

That just about concludes the 'Chemical Techniques' section. As we've seen, some important types of problem can be tackled using them, and if your sample is scarce, all is not lost – it can often be recovered, though this might take some effort on your part. Deuterated samples can be back-exchanged by shaking with an excess of water, trifluoroacetylated samples can be de-acetylated by base hydrolysis, and shift reagents can be removed by chromatography. Now we can move on to consider further (instrumental) elucidation techniques in the next chapter.

8

Further Elucidation Techniques – Part 2

8.1 Introduction

Much of the research in the field of NMR spectroscopy has been in the field of devising new and improved techniques for extracting ever more information from samples. Nowadays, the plethora of available techniques can be daunting for the relative newcomer to NMR. In the following sections, we shall endeavour to guide you through the veritable forest of acronyms by describing the most important and useful techniques and demonstrate how they can be used to solve real-world problems.

Before entering the forest, we would advise you to step back a moment and pause for thought. What information do you require? Is it just a case of an aid to an assignment question, or do you need to discriminate between two or more possible structures? It is important to select the right tool for the job, as some of the experiments we will consider later on can take a significant time to acquire. Doing so will enable you to work more efficiently and have greater confidence in your handiwork.

Many of these instrumental techniques have a two-dimensional counterpart (2-D), which have their own advantages and disadvantages. Rather than treat 2-D spectroscopy as a separate issue, we will include it where appropriate, interleaving it with the corresponding 1-D method. 2-D spectroscopy should perhaps be viewed as an interpretational aid for 1-D spectroscopy, rather than an end itself.

8.2 Spin-Decoupling (Homonuclear, 1-D)

This is probably the oldest of the instrumental techniques but it is still very useful even today. It enables the user to determine which signals in a spectrum are spin-coupled to each other. It can be an extremely useful aid to assignment and can, in some cases, even be used to facilitate conformational studies.

In practice, a powerful secondary radio frequency is centred on the signal of interest while the spectrum is re-acquired. This causes the irradiated proton(s) to become saturated which effectively destroys any spin-coupling from the protons giving rise to this signal. By comparing

Essential Practical NMR for Organic Chemistry, Second Edition. S.A. Richards and J.C. Hollerton.
© 2023 John Wiley & Sons Ltd. Published 2023 by John Wiley & Sons Ltd.

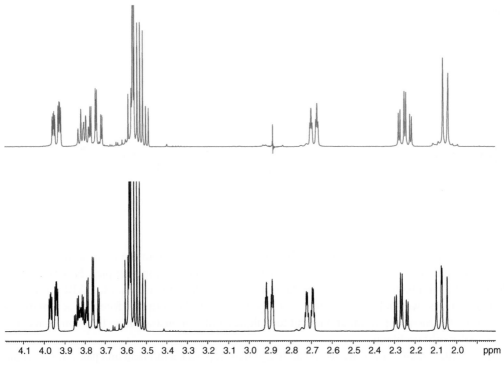

Spectrum 8.1   1-D Spin decoupling experiment (decoupled at 2.9 ppm).

the resultant spectrum with its un-decoupled counterpart, it should be easy to work out which protons couple to the signal of interest. This is demonstrated in the example below. Note that this technique is applicable to both FT and the older CW instruments. The technique is demonstrated in Spectrum 8.1 using *the* morpholine compound…

For convenience and ease of interpretation, it is a good idea to plot the decoupled spectrum above the normal 1-D trace so that you can see at a glance which signals have been decoupled and which have not. The first thing that you'll notice is that the irradiated signal (2.90 ppm) has been obliterated by the decoupler. In our example, the loss of the major coupling from the multiplet at 2.07 ppm, a minor coupling from the multiplet at 2.71 ppm and another from the multiplet at 3.82 ppm are all clearly visible.

1-D decoupling is a very useful tool for unpicking spin systems in this way. You can work your way through your spectrum, decoupling one signal at a time and building up a picture of your structure as you go. Although hardly cutting edge, the 1-D decoupling can offer advantages over the 2-D COSY technique in circumstances where finding an actual value for a coupling is important as well as just establishing connectivity.

8.3   Correlated Spectroscopy (COSY)

Of course, you can find yourself looking at spectra that are complex enough to warrant numerous decoupling experiments for elucidation. In these circumstances, running a single COSY (2-D) experiment as an alternative might well be the answer. A full explanation of the theoretical considerations behind this and other 2-D techniques is well outside the scope of this book but in brief, it works something like this…

Firstly, it is useful to understand what we mean by 1-D and 2-D experiments. If you consider a normal proton spectrum, it is plotted in two dimensions (chemical shift on the 'X' axis and intensity on the 'Y'), so why is it called 1-D? In fact, when NMR started, it wasn't because there was no need to distinguish it from what we now call '2-D'. The number of dimensions that we are talking about are not the dimensions of the number of frequency dimensions that the data set possesses. To try to understand we need to explain the basics of the pulse program. If we take a simple example (for example 1-D proton) we can represent the pulse sequence in Figure 8.1:

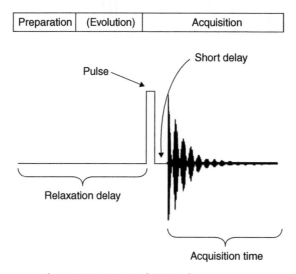

Figure 8.1   A typical 1-D pulse sequence.

This diagram shows that we wait for a certain time (the relaxation delay) and then generate a radiofrequency pulse. We then wait for a short while (to let that intense pulse purge itself from the circuits), switch on the receiver and start receiving the signal. In most experiments we then do it again and again, averaging the spectra that we receive. We generalise these pulse sequences into three components: preparation, evolution and acquisition. In our basic 1-D pulse sequence, there is no 'evolution' bit but this is the key part when we look at 2-D experiments.

What makes 2-D different is that it uses this 'evolution' time to allow something to happen to the spins in the molecule. This can be seen graphically in this simple COSY pulse sequence (Figure 8.2).

In this case we pulse at the beginning of the evolution time and then wait before doing our acquisition pulse. If we vary this wait by incrementing it for each successive cycle, we can change what we see in the FID. This is what generates our second dimension. In the case of the COSY experiment, we allow the coupling information to evolve during this period and then 'read' what has happened to it with the acquisition pulse.

Once we have acquired the data, we have two 'time domains' (one from the normal acquisition time, the other from the incremented delay, hence the data is now '2-D'). As with normal spectra, we need to look at the data in the frequency domain. We do this by Fourier transformation, first in one dimension and then in the other. After the first Fourier transform, we can see a

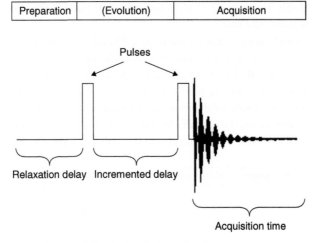

| Preparation | (Evolution) | Acquisition |

Figure 8.2   A simple COSY pulse sequence.

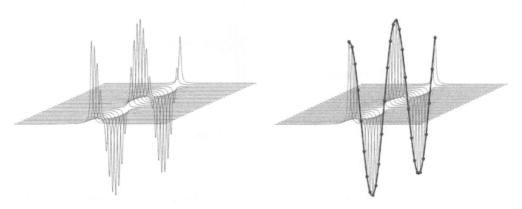

Figure 8.3   Modulation in the second dimension.

number of spectra (each 'increment') where the intensities (or phases) of the peaks are modulated (Figure 8.3).

You can see that (in this case) the modulation follows a cosine relationship and may be Fourier transformed into a second frequency domain, giving us our second dimension.

The resultant data can be portrayed or plotted in one of two different formats. A typical 'stack plot' is shown below (Figure 8.4) and while the intriguing appearance may conjure images of 70's prog rock album covers, stack plots are not in the least user-friendly in terms of interpretation!

For this reason, COSY (and other 2-D spectra) are invariably plotted using a 'map' view or 'contour plot' where contours indicate the intensity of the peaks (Spectrum 8.2 shows a COSY spectrum of the interesting region of *the* morpholine compound). It is worth spending a little time familiarising yourself with the use of a COSY spectrum using this example of a familiar compound. Select, for example, the signal at 2.7 ppm and locate it on the diagonal. Now, using a ruler, project vertically from the diagonal at this point until you connect with a contour. From this contour, project horizontally back to the diagonal. These two signals are spin coupled to

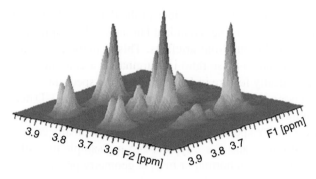

Figure 8.4 A COSY data set.

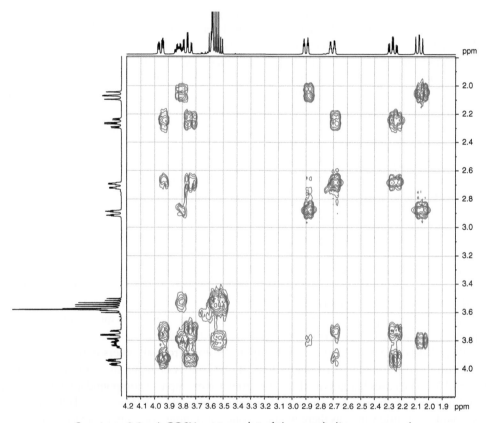

Spectrum 8.2 A COSY contour plot of *the* morpholine compound.

each other. Now return to the peak at 2.7 ppm again and project vertically downwards from it until you encounter two more contours…

It is worth noting that in order to observe all the small couplings, it might be necessary to plot the spectrum with varying intensities. Too low a level will sometimes fail to show all the small couplings while too high a level may cause an unacceptable spread of the diagonal and the stronger correlations.

The diagonal (bottom left to top right) shows the tops of the peaks, as if you were looking down on the peaks of a 1-D spectrum from above. The off-diagonal contours show the couplings between signals and are duplicated on both sides of the diagonal. This might appear strange as half the information portrayed must be redundant but in fact this duplication can be useful as it enables us to tell true correlations from artefacts. This is particularly valuable when dealing with spectra where the signal-to-noise ratio is marginal or a coupling is weak – the coupling contours should always ideally feature in *both* halves of the spectrum.

As you can see, a major advantage of COSY over a conventional 1-D decoupling experiment is that ALL the couplings are displayed on the single plot. On the minus side, it takes a good deal longer to acquire a COSY spectrum as it is made up of typically 256 or 512 individual 1-D spectra, each of at least four scans. For a typical 2–5 mg sample in a 400 MHz spectrometer with an ordinary probe (i.e. 5 mm normal or inverse geometry probe), a high-quality spectrum will take about half an hour to acquire. If your spectrometer is blessed with gradient capability, you can lose the requirement to perform a phase-cycle and acquire a single scan per increment, decreasing the time to around 5 minutes (as long as you have sufficient material).

Another disadvantage is that for solving certain stereochemistry problems, it is necessary to be able to not only establish connectivity but to measure couplings fairly accurately so that the data can be used in conjunction with the Karplus curve. While this is possible using a 'Phase Sensitive' COSY (Note: there are numerous variants of this experiment using various modifications of the pulse sequence, each offering certain advantages/disadvantages), we certainly wouldn't recommend it because of the limited digital resolution available. (Note that in order to avoid collecting gigantic amounts of data, a typical COSY data matrix may be typically 2K in one dimension and 256 points in the other. For a typical 10 ppm sweep width, this means that in a 400 MHz spectrometer, the digital resolution will be at best, $400 \times 10/2048$, or in other words, 2 Hz per point. This would obviously not be good enough to measure couplings to the accepted tenth of a hertz!)

8.4   Total Correlation Spectroscopy (TOCSY) 1- and 2-D

The TOCSY techniques, which come in both 1- and 2-D versions, offer an alternative to 1-D spin-decoupling and COSY methods for establishing through-bond connectivities. The important difference between the two is that TOCSY methods allow easy identification of isolated spin systems. For example, using our trusty morpholine compound once more, you can see that it is possible to identify the $-CH_2-CH_2-$ spin system between the nitrogen and the oxygen atoms, these hetero-atoms, effectively isolating the protons from all others in the molecule.

This ability to discriminate between protons of one spin system and those of another can be very useful in some cases but not in others. Imagine, for example, a compound analogous to our morpholine but with the oxygen and nitrogen replaced by CH_2s. In this case, TOCSY experiments would be of little value, as there would be one continuous coupling pathway, right around the molecule and the resultant TOCSY would look much the same as a corresponding COSY.

In the 1-D experiment, you select a clear (i.e. not overlapped) signal for irradiation and after initiating the appropriate pulse sequence, the resultant spectrum will show only those protons that are in the same 'coupling network' as the selected proton(s). The intensity of the signals produced ultimately dies away with increasing number of bonds from the selected proton(s) but by varying one of the delays in the pulse sequence (the spin-lock pulse), the experiment can be fine-tuned for 'range'. A relatively short spin lock will give rise to shorter range (i.e. weaker or non-existent correlations to distant protons) while a relatively long spin lock will favour long-range correlations though in this case, care must be taken not to damage the probe by pushing too much energy through it.

In the 2-D experiment, as in the COSY, no selection of any signal is required. The sequence is initiated and the data collected.

8.5   The Nuclear Overhauser Effect (NOE) and Associated Techniques

Whereas spin decoupling, COSY and TOCSY techniques are used to establish connectivities between protons *through bond*, techniques that make use of the Nuclear Overhauser Effect (1-D NOE and NOESY, 1-D and 2-D GOESY, 1-D and 2-D ROESY) can establish connectivities *through space*. Before looking at these techniques in detail, it's worth spending a little time considering the NOE phenomenon itself – in a non-mathematical manner, of course!

A working definition of the nuclear Overhauser effect would be, 'A change in the intensity of an NMR signal from a nucleus, observed when a neighbouring nucleus is saturated.' Such changes in intensity may be positive or negative (depending upon how the observation is made, the tumbling rate of the molecule in solution and the frequency of the spectrometer used) and they can be observed in both the homonuclear and the heteronuclear sense. The maximum theoretical magnitude for such effects in steady-state experiments (simple 1-D NOE – difference experiments) is 50% (of the size of the original signal, observed without any secondary irradiation of any neighbouring nuclei) but in reality, they tend to be a lot smaller, usually less than 10% and often as small as 1% but nonetheless, still relevant.

For this reason, they are best observed using a 'difference technique', i.e. a pulse sequence which allows subtraction of two data sets, allowing only differences to be observed and unchanged signals to be edited out of the spectrum. The advantage to this approach should be clear if you consider attempting to observe a change in intensity of 2% in a peak that is 100 mm in height. Would 102 mm look significant? Probably not – but the difference between a peak of 2 mm and no peak at all would be immediately apparent! Note that since you might be looking for an enhancement of less than 2% (i.e. signal intensity of less than 2% of the original spectrum), signal-to-noise ratio may well be an issue and acquiring the data could take a significant time. If you were to investigate half a dozen different sites within a molecule, running the experiments overnight would be advantageous!

In the definition above, the term 'neighbouring nucleus' was used. The Nuclear Overhauser Effect is highly distance dependent – so much so that it falls off with the sixth power of the distance separating the nuclei in question. This very sharp distance dependency makes the effect a very useful tool for probing inter-atomic distances. Two nuclei separated by 3.5 Å should experience the effect between them, but should that distance be 4 Å, they will not. Another important point to bear in mind is that in marked contrast to spin-coupling, though proton (x) gives an NOE to another nucleus (y), there is no guarantee that (y) will give an NOE back to (x). This is because (y) might have more favourable relaxation pathways available to it.

The ability to devise experiments that can make use of the Nuclear Overhauser Effect gives us massively powerful tools which can be used to crack all manner of problems. For example, they can be used in the more trivial sense, as an assignment aid and to tackle problems of positional isomerism. But the area where NOE experiments really come into their own by offering information that no other NMR techniques can offer, is in the field of stereochemistry. Is this group up or down? Could this centre have epimerised? An NOE experiment could be just what you need...

In the basic 1-D NOE experiment, the spectrometer collects two sets of FIDs, one with a second RF source centred on the signal to be examined and a second set with the same RF source centred on a blank part of the spectrum. After a suitable number of both sets of scans have been acquired (an equal number of both!), the two sets are subtracted from each other to leave a resultant spectrum which should only show signals of protons that have undergone enhancement because they were within approx. 3.5 Å of the proton(s) that was irradiated. Note that during the acquisition pulse, the decoupler is switched off and so the enhanced signals retain any coupling they may have. Note also that subtraction may not be perfect and that the enhanced spectrum may contain a few subtraction artefacts. These can usually be easily distinguished from genuine enhanced peaks as they cannot be phased, have intensity above and below the baseline and usually have no net integration associated with them. Note also that it is advisable to run NOE experiments with the sample *not* spinning. This helps minimise subtraction artefacts by broadening the peaks very slightly.

As an example of how useful an NOE experiment can be, consider the structures below:

These two compounds would give very similar proton (and carbon) spectra and though an educated estimate of the –CH$_2$– and –CH$_3$ chemical shifts would give a good indication of identity *if both compounds were available*, we would never entertain such liberties if we had only one of the compounds in isolation. (Note that the chemical shifts of these substituents would be expected to be at slightly lower field when they are in the alpha positions. This is because alpha substituents are deshielded by two aromatic rings whereas those in beta environments are subjected to deshielding by only one of the aromatic rings. The differences involved would only be of the order of 0.2 ppm.) An appropriate NOE experiment, however, removes all speculation and in combination with relevant decoupling/COSY, rapidly yields a full and unambiguous assignment of the molecule.

In this example, both the –CH$_2$– and the –CH$_3$ would be excellent targets for irradiation and we would recommend making use of both of them. A brief inspection of the 1-D spectrum (Spectrum 8.3) is enough to confirm that the compound does have both substituents on one of the rings as four protons can easily be observed as one continuous spin system (8.13, 7.85, 7.6 and 7.48 ppm) while the remaining two signals are a pair of ortho-coupled doublets at 7.77 and 7.34 ppm. This proves that both substituents are not only on the same ring but also that they must be either ortho- or para- to each other. The first NOE experiment (Spectrum 8.4) in which the –CH$_2$–Cl protons are irradiated gives a clear enhancement of the broad doublet at 8.13 ppm. The –CH$_3$ protons are also enhanced which shows that these substituents are ortho to each other. (Note that the NOE trace is plotted in red above the standard 1-D plot and on the same scale for convenience.) The enhancement of the broad doublet at 8.13 ppm is entirely consistent with structure 'A' above.

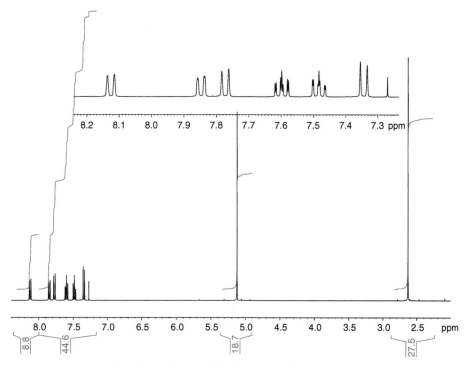

Spectrum 8.3 Naphthalene substituted with –CH₃ and –CH₂-Cl groups and expansion.

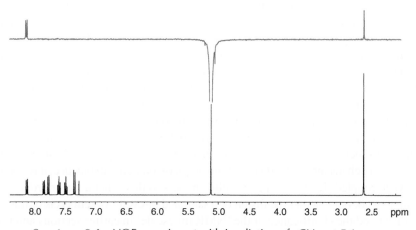

Spectrum 8.4 NOE experiment with irradiation of –CH₂– at 5.1 ppm.

The second NOE targeting the methyl group (Spectrum 8.5) shows an enhancement of the doublet at 7.34 ppm which underpins the structure which is shown below with the enhancements depicted. The differentiation of the two structures is therefore unambiguous and the correct structure with enhancements is shown below.

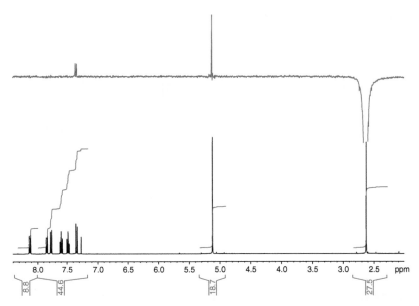

Spectrum 8.5   NOE experiment with irradiation of –CH₃ at 2.63 ppm.

There are a number of pitfalls waiting for the unwary when setting up and interpreting the results of NOE experiments. For example, the signal that is being irradiated should not be too close to any other signal in the spectrum. This is because there is a danger of 'spill over' from the decoupler signal so that you might inadvertently saturate a nearby peak which could of course give rise to completely bogus enhancements. (Note that this is only a potential problem in the 1-D techniques where selective irradiation of a specific signal is used.) In the 1-D experiments, the irradiated signal always shows the opposite phase to the enhanced signals and should be plotted so that it is negative. If, during the phasing of your NOE spectrum, any other signal which is close to your target signal phases negatively, then be advised that it has been at least partially saturated and spurious enhancements may be present!

Enhancements between signals that are strongly spin-coupled to each other are best ignored as they are prone to another competing phenomenon, that of Selective Population Transfer (SPT). This makes it difficult to decide if any observed enhancement is down to a genuine NOE, or is merely an SPT. SPT signals are characterised by their inability to phase properly. In practice, it is not often that an NOE between coupled signals would be useful anyway so this is not a major problem…unless you are trying to work out whether the fusion of a saturated bi-cyclic system is cis or trans…

We have discussed the significance of the 'NMR timescale' in earlier sections and it is worth knowing that the 'NOE timescale' is somewhat longer and that this can have consequences for NOE experiments in molecules that have dynamic processes taking place within them. To give a more specific example, consider the isomers below:

Differentiating these two compounds, particularly in isolation, would not be easy by proton NMR. The temptation to irradiate the –OH should be resisted (Note: The irradiation of exchangeable protons in NOE experiments is not generally recommended, even if they give rise to sharp peaks) as these compounds can undergo tautomerism and exist in the forms:

This phenomenon could give no end of potential problems with an NOE experiment. The molecules may exist in solution, predominantly as either hydroxy pyridines or as pyridones, depending to some extent on solute concentration, choice of solvent, its water content, pH and temperature. Irradiation of the exchangeable signal would therefore be an uncertain proposition as you could not be sure what exactly you were irradiating! Furthermore, it is quite possible that the two tautomers could both exist in solution simultaneously. Tautomerism is *generally* fast on the NMR timescale, i.e. we *usually* see only one set of signals that represent the average contributions of the chemical shifts of both tautomers. During an NOE experiment, it would be likely that both forms would effectively undergo irradiation because an irradiated –OH, undergoing chemical exchange to become an NH, *takes its irradiation with it (and vice versa of course)!* You would effectively be irradiating both sites at once. Should the proton transfer process turn out to be slow on the NMR timescale, i.e. you observe two distinct sets of signals for the two different tautomers, it would still be relatively fast on the NOE timescale (because the experiment requires a low power irradiation of the signal under investigation which generally lasts for at least 1 second. During this time, chemical exchange inevitably occurs) and both exchangeable sites would still be irradiated. This would obviously give rise to useless data and meaningless results.

This exchange process can also be a problem where the water in a solvent becomes unintentionally irradiated during an NOE experiment because the protons of the water are in constant chemical exchange with all exchangeable protons in the molecule being studied. Consider for example, the following hypothetical problem. You wish to distinguish between the two compounds shown below:

A reasonable strategy might be to positively identify the proton ortho- to the –OCH$_3$ group by means of an NOE experiment and then use this proton as a further probe in a second NOE experiment to see if it enhanced either the NH or possibly acetyl methyl in one isomer, or the peri- aryl proton in the other. A problem could arise here, using DMSO as a solvent perhaps, if irradiation of the –OCH_3 accidentally irradiated the water present in the solvent. Don't forget – you only have to catch the edge of the peak to saturate it. The (irradiated) water could chemically exchange with the NH, passing irradiation on to it and in so doing, initiate a bogus enhancement from the NH to the proton peri- to it which would beg the question: 'Did the enhancement come from the –OCH_3 or from the NH (relayed from the water)?' In this case, a better method might be to work from the aromatic signals of the AB pair to establish connectivity *to* the –OCH_3 rather than *from* it.

These are just a few examples of what could go wrong with an NOE experiment. NOE experiments are not 'boring' and 'all the same' as a chemist acquaintance once famously remarked. Quite the contrary in fact. Designing sensible experiments to make use of the Nuclear Overhauser Effect and dovetailing the results with other NMR data can be quite challenging – and rewarding when you finally pull all the threads together to produce a sensible picture of a problem molecule.

At the beginning of this section, we listed the various experiments that are available which make use of the Nuclear Overhauser Effect but as yet, we have made no attempt to indicate the pros and cons of each of these and under what circumstances one may be preferable over another. It is virtually impossible to give cast iron advice regarding the selection of one NOE experiment over another as the decision has to be based on a huge number of considerations, and on the instrumentation and software available to you. Having said that, we shall now attempt to establish some broad guidelines…

Perhaps the first decision to be made is whether to select a 1-D or a 2-D technique. Note that in all the 2-D NOE experiments, the off-diagonal-peaks contours represent NOE connections between signals and are displayed on each side of the diagonal in exactly the same way that coupling connectivities are displayed in COSY spectra. Both have their advantages and disadvantages. If you are working on a relatively simple problem such as that of the –CH_2–Cl and –CH_3 groups on the naphthalene which we considered earlier, then a 1-D approach would be preferable since the problem could be cracked with a single NOE experiment, or two at the most and this could be achieved more quickly than by running a 2-D experiment. The simple 1-D NOE is a robust and trustworthy tool. For more complex problems, where you might benefit from having NOE data from multiple sites, a 2-D technique might be preferable as it should give you all the available NOE information about the molecule in one spectrum. Both 1- and 2-D techniques can suffer from artefacts (features in the spectra that are not genuine NOE signals). We have already mentioned subtraction errors in the basic 1-D method but perhaps some of the artefacts that can occur in 2-D spectra can be more serious. For example, 'T1 noise', which manifests itself as a streak of cross-peaks running down the spectrum in a line with any strong peaks on the diagonal, can cause problems. This type of streak can obscure genuine correlations. The severity of T1 noise is an instrumental factor that is related to RF stability and thus varies from instrument to instrument.

One potential problem that can occur with slightly larger molecules (typically of m.w. > 600) is that the NOE response in both NOE and 2-D (NOESY) experiments is related to the tumbling rate of molecules in solution. The larger the molecule, the slower it will tumble and at a certain point, all expected enhancements will be nullified. This null-point depends not only on the tumbling rate (and therefore the size, or more accurately, the shape of the molecule) but also on the field strength of the spectrometer being used. A molecule giving positive NOEs in a 400 MHz instrument may well not give NOEs in a 600 MHz machine – or maybe it will give negative NOEs…

In order to combat this, the ROESY techniques (Rotating frame Overhauser Effect Spectroscopy) can be employed. An in-depth discussion of how this technique works is outside the remit of this book but suffice to say, in the ROESY methods (1- and 2-D), NOE data is acquired as if in a weak magnetic field rather than in a large, static magnetic field and this assures that all NOEs are present and positive, irrespective of tumbling rate and magnet size. It is possible that some TOCSY correlations can break through in ROESY spectra but these will have opposite phase to the genuine ROESY correlations and so should therefore not be a problem – unless they should overlap accidentally with them. A 2-D ROESY spectrum of the naphthalene compound is shown below (Spectrum 8.6).

A comparison between the 1- and 2-D data shown for this compound is interesting. As we have said, the 2-D ROESY does offer the advantage of displaying all enhancements occurring in the molecule simultaneously but against that, the data is probably more prone to artefacts than the corresponding 1-D technique. This can be particularly apparent in cases where the transmitter offset frequency (which defines the centre of the sweep width) happens to coincide with a signal in your spectrum! In terms of making optimum use of spectrometer time, the 1-D experiment would be the preferred choice in cases where you only have a few 'target' signals to irradiate while the 2-D method might be the best choice in cases where you need to look at four or more signals. The 1-D experiment also offers another advantage in that the enhanced signal is 'reconstructed'. This can be very useful if this signal is overlapped with other signals which do not enhance, as it provides us with a method of extracting coupling information not available in the standard 1-D spectrum.

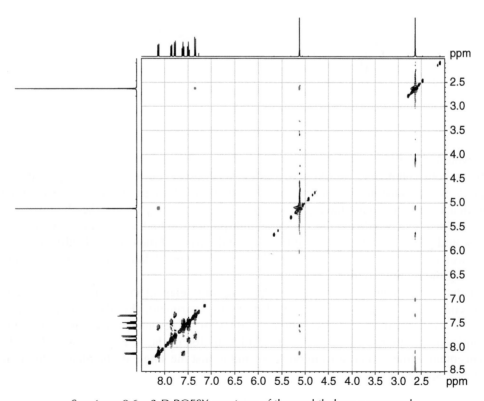

Spectrum 8.6   2-D ROESY spectrum of the naphthalene compound.

At the beginning of this section, we listed 1- and 2-D GOESY as alternative methods of collecting NOE data. This technique (Gradient enhanced Overhauser Effect Spectroscopy) is broadly similar to conventional NOE in terms of the results you achieve. In the 1-D case, there are no subtraction artefacts since the subtraction of data is handled by a phase cycle. Viewed pragmatically, GOESY spectra are generally cleaner but offer no notable advantage in terms of signal/noise. It would seem that the conventional NOE method might also be somewhat more robust – we have seen examples of problems that have not given an expected enhancement in a GOESY experiment but have given perfectly acceptable results in a conventional NOE experiment.

So to sum up, if you have a small molecule, a straightforward issue to resolve and a typical 250/400 MHz instrument at your disposal, use an ordinary 1-D NOE. If you have a more complex problem involving multiple sets of NOE data to consider, go for a 2-D method and if you have a larger molecule and a more powerful spectrometer, go for a ROESY option.

We have concentrated on the proton–proton, homo-nuclear NOE experiments in this section but the potential use of analogous hetero-nuclear experiments should not be overlooked, if you have the appropriate hardware available to you. The ^{19}F–proton NOE experiment, for example, can be very useful in certain situations as demonstrated in the following example. You have one of two possible positional isomers:

How would you differentiate between them? This problem is not a good one for proton NMR as both compounds would give similar spectra (if you had both compounds, you might draw a reasonable conclusion on the basis of the –CF$_3$ group's ortho- deshielding). Note that both compounds have protons that are ortho- and para- to the shielding –OH group and that they would exhibit the same multiplicities in both compounds. ^{13}C spectroscopy would give a good indication of identity on the basis of the chemical shifts of several of the aromatic carbons – but you would need access to a good data base to have confidence in solving the problem in this way.

But the most unambiguous and arguably the most elegant confirmation of structure would come in the shape of a hetero-nuclear NOE experiment. (Firstly, you have to run a quick ^{19}F spectrum in order to determine the relevant ^{19}F resonance frequency and set the decoupler in the fluorine domain, of course.) Irradiation of the –CF$_3$ group would yield an enhancement of the two protons in one case and to just the single deshielded proton in the other. See Spectrum 8.7.

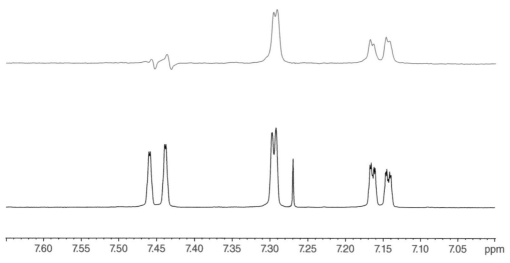

Spectrum 8.7   An NOE experiment with irradiation of the –CF$_3$ group at –62.85 ppm in the ^{19}F domain.

The enhancement of the two protons as shown below clearly defines the isomer.

Note that in cases where ^{19}F–^{1}H NOE experiments are attempted in molecules where fluorine is spin-coupled to any of the protons within NOE range, SPT effects can be expected as described earlier!

There are several other extremely useful techniques for the elucidation of structures that we use regularly but since these all make use of ^{13}C data, we'd better start a new chapter.

9

Carbon-13 NMR Spectroscopy

9.1   General Principles and 1-D ^{13}C

^{13}C NMR gives us another vast area of opportunity for structural elucidation and is incredibly useful in many cases where compounds contain relatively few protons, or where those that are available are not particularly diagnostic with respect to the proposed structures. Before we delve into any detail, there are certain general observations which we need to make regarding ^{13}C NMR and the fundamental differences that exist between it, and 1H NMR.

For a start, we must be mindful of the fact that ^{13}C is only present as 1.1% of the total carbon content of any organic compound. This, in combination with an inherently less sensitive nucleus, means that signal-to-noise issues will always be a major consideration in the acquisition of ^{13}C spectra – particularly 1-D ^{13}C spectra which we will restrict the discussion to for the moment. (Note that the *overall* sensitivity of ^{13}C, probe issues aside, is only about 0.28% that of proton because the nucleus resonates at a far lower frequency – in a 400 MHz instrument, ^{13}C nuclei resonate at around 100 MHz.) So it takes a great deal longer to acquire ^{13}C spectra than it does proton spectra. More material is obviously an advantage but can in no way make up for a 350-fold inherent signal-to-noise deficiency!

Another important aspect of ^{13}C NMR is that the signals are never normally integrated. The reason for this is that some carbon signals have quite long relaxation times. In order to make NMR signals quantitative, acquisition must allow for a relaxation delay (delay period between acquisition pulses) of at least five times the duration of the slowest relaxing nuclei in the compound being considered. With relaxation times of the order of 10–20 seconds, it is therefore obvious why we cannot obtain quantitative ^{13}C data. The inherent insensitivity of the ^{13}C nucleus often demands thousands of scans to achieve acceptable signal/noise so we can ill afford 100 second relaxation delays between pulses! The only thing that we can say is that methine, methylene and methyl carbons *generally* appear to be more intense than quaternary carbons. (See below for explanation.)

Yet another significant difference with ^{13}C NMR is that we do not observe coupling between neighbouring nuclei as we do in proton NMR. This is not by virtue of any decoupling technology – it is purely a matter of statistics. As the natural abundance of ^{13}C is only 1.1%, the chances

Essential Practical NMR for Organic Chemistry, Second Edition. S.A. Richards and J.C. Hollerton.
© 2023 John Wiley & Sons Ltd. Published 2023 by John Wiley & Sons Ltd.

of having two ^{13}C atoms sitting next to each other is statistically small (it would occur in only 1.1% of the molecules in fact) and so ^{13}C–^{13}C coupling just isn't an issue.

^{1}H–^{13}C coupling, however, *would* be a serious issue – if it were allowed to occur. 1-D ^{13}C spectra are always normally acquired with full proton decoupling. There are a number of good reasons for this. Firstly, the already meagre signal-to-noise would be further eroded by splitting the signal intensities into doublets, triplets, etc. Furthermore, identifying individual signals would be extremely difficult in compounds having many carbons in a similar chemical environment – particularly in view of the large couplings that exist between protons and ^{13}C nuclei. The potential overlap of signals would make spectra horrifically complex.

Another good reason for fully decoupling protons from ^{13}C is that the ^{13}C sensitivity, to some extent, benefits from Overhauser enhancement (from proton to ^{13}C which comes about as a result of decoupling the protons). This explains why quaternary carbons appear less intense than those attached to protons – they lack the Overhauser enhancement of the directly bonded proton.

So far, it might seem that ^{13}C spectroscopy is just a long list of disadvantages. Here we have a technique that is extremely insensitive and thus time-consuming to acquire. It is largely non-quantitative, since we can't integrate the signals and, to gild the lily, we can't relate carbon-to-carbon by means of spin coupling as we have no coupling information to assist us in our assignments. Just about the only commodity we have left at our disposal is the chemical shift – but how do you go about interpreting a spectrum that is composed entirely of singlets? We will explore this a bit later. (Note that although all ^{1}H–^{13}C couplings are decoupled, couplings between ^{13}C and other heteroatoms such as fluorine and phosphorus will *not* be decoupled and splitting of ^{13}C signals will be observed in molecules where these heteroatoms are found in environments that are conducive to coupling.)

So if this all sounds a bit bleak, what's the *good* news? Well, strangely, there is quite a lot. For a start, let's not forget that had the ^{13}C nucleus been the predominant carbon isotope, the development of the whole NMR technique itself would have been held back massively and possibly even totally overlooked as proton spectra would have been too complex to interpret. Whimsical speculation aside, chemical shift prediction is far more reliable for ^{13}C than it is for proton NMR and there are chemical shift databases available to help you that are actually very useful (see Chapter 15). This is because ^{13}C shifts are less prone to the effects of molecular anisotropy than proton shifts as carbon atoms are more internal to a molecule than the protons and also because as the carbon chemical shifts are spread across approximately 200 ppm of the field (as opposed to the 13-odd ppm of the proton spectrum), the effects are proportionately less dramatic. This large range of chemical shifts also means that it is *relatively* unlikely that two ^{13}C nuclei are exactly coincident, though it does happen.

Other good news comes in the shape of the ^{13}C nucleus having a spin quantum number of ½. This means that ^{13}C signals are generally sharp as there are no line-broadening quadrupolar relaxation issues to worry about and we don't have to deal with any strange multiplicities.

So to a large extent, 1-D ^{13}C NMR interpretation is a case of matching observed singlets to predicted chemical shifts. These predictions can be made by reference to one of the commercially available databases that we've mentioned, or it can be done the hard way – by a combination of looking up reference spectra of relevant analogues and using tables to predict the shifts of specific parts of your molecule (e.g. aromatic carbons). We have included some useful ^{13}C shift data at the end of the chapter but it is, by necessity, very limited.

^{13}C prediction software is certainly the preferred option but it should always be used with circumspection. It generally works by using a combination of library data to generate an estimate of the chemical shifts of all the carbons in your proposed structure but it is inevitable that these estimates will be prone to error. It is important to realise that some shift estimates will be far better than others – even within the same molecule. It is also important to note that while these packages may give a measure of confidence with each prediction, these limits must be viewed critically as they may be either unduly pessimistic or (worse) unduly optimistic. We would always recommend that if your prediction software allows you to browse the *actual* compounds used in the predictions, you do so! This will enable you to 'personalise' the predictions to some extent as you will be able to lean more towards the shifts of the compounds in the database that are more similar to your proposed structure. For example, if you are working with steroids and you are trying to predict the shift of a certain carbon in your molecule, it would be wise to pay more heed to the shifts of carbons in similar environments *in other steroids* as opposed to analogous carbons in completely different types of molecule.

From a purely pragmatic point of view, and some purists may take issue with this, it is perhaps not essential that you unambiguously assign every carbon to a *specific* peak as this can be virtually impossible in cases where there are many carbons with similar shifts and all you have to guide you is a mediocre prediction. What *is* important is that the total number of peaks observed match the number of carbons in your proposed structure and that all their chemical shifts are at least plausible. We shall see presently that there are other tools available which can be used to yield unambiguous assignments in many cases. Consider the carbon spectrum of our familiar morpholine compound, Spectrum 9.1, which demonstrates this point. The chemical shifts of the two carbons in the morpholine ring next to oxygen are pretty close. So too are the carbons next to the nitrogen…

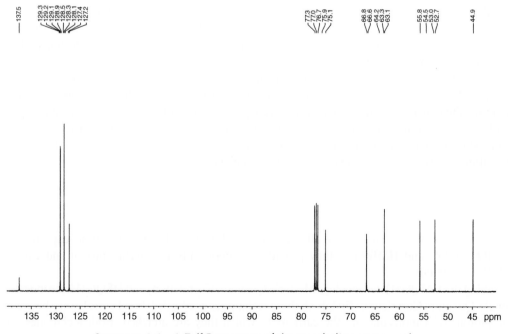

Spectrum 9.1   1-D ^{13}C spectrum of *the* morpholine compound.

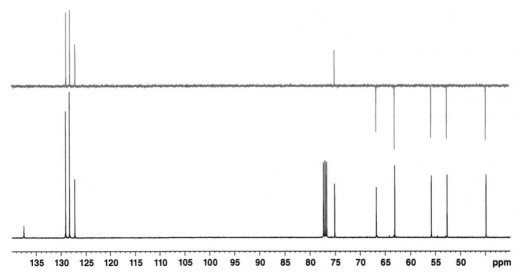

Spectrum 9.2   1-D ^{13}C spectrum of *the* morpholine compound with DEPT-135 plotted above it.

The first of these tools is the DEPT pulse sequence (Distortionless Enhancement by Polarisation Transfer). There are a number of versions of this experiment which can be very useful for distinguishing the different types of carbons within a molecule. Of these, we have found the DEPT 135 sequence to be the most useful. In this experiment, the quaternary carbons are edited out of the spectrum altogether. Methyl and methine protons naturally phase at 180 degrees relative to the methylene carbons and the spectra are usually plotted with methyls and methines positive. (Note that should you encounter a signal that you cannot confidently assign to either a methyl or methine carbon, the DEPT 90 sequence may well be of use as it differentiates these carbons – methines appear positive and methyls are edited out of the spectrum but this technique can be considered obsolete if you have access to any of the 2-D proton–carbon correlated experiments discussed in the next section.)

This is demonstrated once more with our familiar morpholine compound in Spectrum 9.2. The DEPT sequences are of course, still relatively insensitive, though they are probably a little more sensitive than the standard 1-D, fully decoupled ^{13}C spectrum. We find it convenient, particularly with complex molecules, to combine the 1-D ^{13}C spectrum with the DEPT-135 spectrum, which is plotted above it at the same expansion, of course! This enables you to differentiate the different types of carbon in your spectrum at a glance.

9.2   2-D Proton–Carbon (Single Bond) Correlated Spectroscopy

The most powerful techniques of all are undoubtedly, the 2-D proton–carbon experiments (HMQC/HSQC and HMBC) as they provide an opportunity to dovetail proton and carbon NMR data directly.

Taking the HMQC (Heteronuclear Multiple Quantum Correlation) and HSQC (Heteronuclear Single-Quantum Correlation) first, both these techniques establish 1-bond correlations between the protons of a molecule and the carbons to which they are attached. Both techniques are considerably more sensitive than a 1-D ^{13}C spectrum, which might seem strange when you consider that the whole 2-D matrix is composed of a considerable number of ^{13}C spectra. The secret of the superior signal/noise performance of these methods lies in the fact that they are both 'Indirect Detection' techniques. This means that the carbon signals are detected (indirectly) by the transfer of their magnetisation to the much more sensitive protons! A typical data matrix for an HMQC or HSQC might be composed of 256 increments in the carbon domain, each of 2k points in the proton domain. For a 5–10 mg sample of typical 200–400 molecular weight, reasonable signal-to-noise could be achieved with about 16–32 scans per increment in a 400 MHz instrument which means that you could easily achieve a good-quality spectrum in about 1–2 hours.

In terms of choosing between the two, bear in mind that the choice available on the spectrometer you use may well be limited by the hardware itself. Historically, the HMQC looked like the better bet at first, as it was more robust. The HSQC technique was fine – but the large number of 180 degree pulses in the sequence, require accurate pulse calibration if severe cumulative errors are to be avoided. In other words, if the probe tuning was not optimised, you could expect very poor signal-to-noise or even no signal at all. Probe tuning and matching is not the sort of thing you can reasonably expect the average 'walk-up' user to get involved with and for this reason, HSQC was a non-starter. HMQC was the way to go but times and hardware move on and nowadays, most modern instruments *can* deal with HSQC routinely without the need for any poking around under the magnet with a non-magnetic tweaking stick!

The two developments responsible for this are 'automatic probe tuning and matching' and 'adiabatic pulses'. Automatic probe tuning and matching enables optimal probe tuning to be achieved for every sample in an automated run, regardless of solvent. Adiabatic pulses solve the problem in a different way – by removing the need for accurate pulse calibration. Solving this problem enables us to routinely enjoy the benefits of HSQC over HMQC which include fewer spectral artefacts and slightly better resolution in the carbon domain.

So to sum up, if you have the luxury of modern equipment with all the go-faster boxes at your disposal, go for HSQC. If you are stuck with an older instrument and you're not keen on grovelling around under the magnet, an HMQC is for you.

Our preferred experiment of this type is the so called 'DEPT-edited HSQC' which is both relatively artefact-free and sensitive. It also has one other major advantage up its sleeve. This experiment is not an 'absolute value' technique like most of the others, but it allows for discrimination between different types of carbons. Methyl and methine carbons give cross-peaks that are phased opposite to the methylene carbons and so the results are best plotted on a colour plotter which can portray this clearly by plotting positive and negative cross-peaks in different colours.

A brief note on the phasing of the DEPT-edited HSQC spectra – because the technique is 'phase sensitive' (as opposed to 'absolute value'), these spectra require phasing. This is usually done under automation in walk-up systems and usually done well. (Note that phasing has to be performed in *both* dimensions.) Sometimes, you may find a signal at one end of your spectrum which is clearly not phased, despite the fact that all the neighbouring signals appear perfectly phased. The likely reason for this will be that the size of the ^{1}H–^{13}C coupling for the carbon in question is abnormally large or small and there is not much that can be done about it. (Attempts to phase such signals correctly will result in the phasing of all the other signals suffering!) Instruments are typically set up to give a maximum sensitivity for couplings of around 145 Hz. This figure is a compromise between the generally smaller couplings found in alkyl systems and the slightly larger ones encountered in aryl systems. The larger the deviations of the 1-bond ^{1}H–^{13}C couplings from this value, the greater will be the shimming inaccuracies encountered. Typical problem carbons are those of the nitrogen-bearing heterocycles where couplings approaching 200 Hz are quite common.

Interpreting HMQC/HSQC spectra is relatively straightforward as you can see from the HSQC spectrum of *the* morpholine compound (Spectrum 9.3). Basically, it's a case of lining up the proton signal with the contour, and reading off the ^{13}C chemical shift. The technique is extremely powerful – particularly when used in combination with HMBC as we'll see later. In examples like this one, it enables you to identify geminal pairs of protons at a glance as you can see which protons are attached to the same carbons. As a philosophical aside, we should always be on our guard against using proton data to 'hammer' the ^{13}C data to fit. Although the 2-D techniques tie the data sets together, we must still interrogate them separately. In other words, if a correlation flags up an implausible shift in one domain or the other, the whole structure should be re-considered…

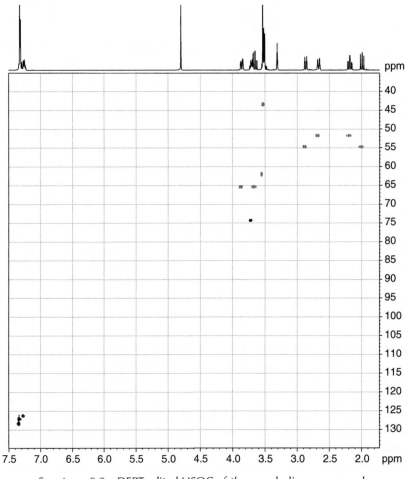

Spectrum 9.3   DEPT-edited HSQC of *the* morpholine compound.

HMQC/HSQC spectra can be extremely useful in resolving problems where there is a significant carbon chemical shift precedent that could be used to support one putative structure over another – for example, in dealing with cases of O- versus N-alkylation. Take for example the two methylated indoles below:

How could you be certain of which site had alkylated? Any judgement based on proton chemical shift would be foolhardy as there would be little to choose between them. (Note that in molecules where the lone-pair of electrons on a nitrogen are effectively 'removed' from the nitrogen for whatever reason – in this case, by donation into the aromatic ring – the nitrogen becomes more electron-deficient and thus more 'oxygen-like'. The chemical shift of alkyl groups substituted onto such nitrogens therefore become very similar to those of analogous O-alkylated compounds and distinction between them on the basis of proton chemical shift becomes unreliable). The methyl groups in both molecules might be expected to give Overhauser enhancements to their nearest aryl protons but in order to make use of this, you would have to be absolutely certain of the assignment of the aryl protons themselves! ^{13}C data would be unambiguous here. A methyl singlet with a carbon shift in the range, 35–45 ppm and you are looking at N-methylation. If the carbon shift of the methyl is in the region of 55–65 ppm, it's the oxygen that has been methylated.

If interpreting the single-bond correlation experiments is easy, the multiple bond experiment (HMBC) can be considerably less so…

9.3 2-D Proton–Carbon (Multiple Bond) Correlated Spectroscopy

Potentially even more useful is the HMBC (Heteronuclear Multiple-Bond Correlation) experiment. In this experiment, correlations are obtained between carbon atoms and protons that are separated by two and three bonds. Of course, the actual number of bonds separating the protons from the carbons is something of a red herring. What the spectrometer records are carbon–proton contours for carbons that have protons couplings of specified magnitude. The sensitivity of the spectrometer to various sizes of proton–carbon coupling is controlled by one of the delays in the HMBC pulse sequence. This delay is selected on the basis of 1/2J, where J is the coupling you wish to optimise for. A proton–carbon coupling of 10 Hz is a fairly typical value for the experiment, and thus the relevant delay would be set at 1/2 × 10, or 0.05 seconds. This means that the spectrometer sensitivity would be optimised for carbons with proton couplings of around 10 Hz. It does not mean that it will not detect carbons with smaller or larger proton-couplings, just that the response shown will not be as intense.

In practice, 3-bond proton couplings tend to be nearer to this value than are the 2-bond couplings and, for this reason, the HMBC sequence is usually more sensitive to 3-bond than to 2-bond correlations. This has of course to be viewed within the context of the overall signal/noise for the experiment. If the signal-to-noise for the whole experiment is less than excellent, it is quite possible for some 2-bond correlations to slip through the net altogether. If you are wondering why the value of 1/2J is not used to even-up the response to 2- and 3-bond correlations, there are two important factors to consider. If this value was optimised for, say, 5 Hz proton couplings, then the spectra we would obtain would be further greatly complicated by 4-bond couplings which would start to come through, since the J-values for some 4-bond couplings are comparable with 2-bond values. Furthermore, as the value for J gets smaller, so the optimal delay required gets longer so that more and more signal gets lost to relaxation prior to acquisition and overall sensitivity for the experiment is lost. This incidentally explains why the technique is not as sensitive as HSQC in the first place (1-bond proton–carbon couplings are typically around 150 Hz, so the delay is extremely short and very little signal is lost).

So, putting it bluntly, HMBC spectra are more difficult to un-pick because there will inevitably be far more correlations recorded than in the corresponding HMQC/HSQC. The problems do not end there, however. For example, it is not immediately obvious, by inspection, which are the 2-bond and which are the 3-bond correlations. This has to be reasoned out within the context of whatever molecule you are dealing with. Furthermore, while most 4-bond proton–carbon couplings are less than 2 Hz, some are not, allowing unwanted 4-bond correlations we've mentioned, through into our spectra even when we've optimised for 10 Hz couplings! This can be a problem particularly in the case of aromatic, heterocyclic and conjugated compounds where signal-to-noise is good. These need to be identified for what they are as soon as possible or they will cause confusion!

Unfortunately, the complexity does not stop there. 1-bond couplings can also come through in the HMBC experiment, despite filters used to block them. This can be seen in our HMBC spectrum of *the* morpholine compound (Spectrum 9.4) with reference to one of the aromatic signals at 126 ppm. 1-bond correlations are characterised by a pair of contours that are symmetrically displayed on either side of the 1-D (proton) projection they relate to, the separation between them giving the proton-carbon coupling, of course. While this is generally not a problem for obvious isolated singlets, it certainly can be a problem in the crowded aromatic region of the spectrum where chemical shifts are relatively tightly packed in both proton and carbon dimensions. Problems can arise where 1-bond contours fall in positions where they line up exactly with peaks from the 1-D proton projection, giving rise to potentially very confusing bogus 'correlations'.

Another feature that is worth being aware of is the so-called 'auto-correlation' phenomenon. These can be observed in molecules which contain moieties such as –N-dimethyl or isopropyl groups. Such groups can give the initially confusing arrangement of three contours in a row. In such cases, the two outer contours are the 1-bond correlation, and, the centre correlation, the genuine 3-bond HMBC correlation from the protons of one of the methyl groups to the carbon of the *other* methyl group. Once this pattern has been noted, you will recognise it easily and even make use of it as a quick identifier for these groups.

If all these cautionary notes make the technique sound like a complex nightmare, we're not done just yet… Just as an unwanted 4-bond correlation can come through to muddy the water and a 2-bond coupling can fail to materialise, so too can a 3-bond coupling fail to register for exactly the same reason – the size of the proton–carbon coupling may be too far from the optimised value to give a sufficient response to be recorded. There can be two possible reasons for this.

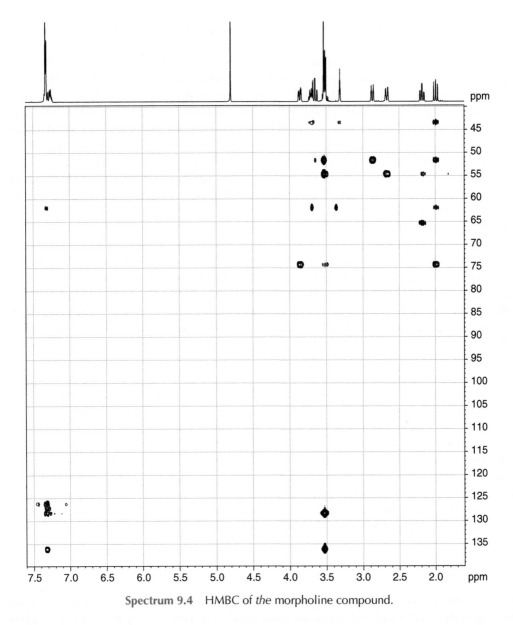

Spectrum 9.4 HMBC of *the* morpholine compound.

Firstly, it can just be a question of local electron distribution giving rise to an abnormal value for the 3-bond proton–carbon coupling. One that springs to mind is the lack of a correlation often observed between the 3′ proton and the 4′ carbon in indoles:

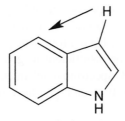

Expected 3-bond correlation often not observed.

Another reason for not observing expected 3-bond couplings relates to the Karplus equation which we discussed at length in the section dealing with carbocyclic compounds. Just as the size of proton–proton vicinal couplings are dependent on the dihedral angle between them, so too are proton–carbon couplings. You can come across molecules where the relevant angle suppresses the coupling and hence a 3-bond proton–carbon coupling can fail to show.

Our advice is that if it is vital to establish such a connectivity, then re-run the experiment optimising for a smaller coupling value (e.g. 5 Hz instead of 10). Yes, you will take a hit on signal-to-noise and spectral complexity may well increase as 4-bond couplings start to come through but if you are chasing down one specific coupling, then these things don't really matter.

HMBC experiments are not limited to proton–carbon interaction. With suitable hardware, it is possible to acquire ^1H–^{15}N spectra which can be extremely useful for confirming the identity of nitrogen-containing heterocyclic compounds. The sensitivity of this technique is very low, probably only about a tenth of the ^1H–^{13}C technique but sometimes it can provide that extra, vital piece of the jigsaw. We have provided some basic but useful ^{15}N shift data in the next chapter.

After digesting this information and noting the many benefits of the 2-D proton–carbon techniques (providing the pitfalls and complexity of the HMBC technique haven't put you off the idea!) you might be wondering why anybody would ever bother acquiring a simple 1-D ^{13}C spectrum any more. Well, there are two good reasons that spring to mind… Firstly, it is quite possible to encounter a molecule with no protons within a 3-bond range of one or more of its carbon atoms. Such carbons will be 'invisible' to the HMBC technique and will only be visible in a 1-D spectrum.

Secondly, the resolution achieved in a 2-D experiment, particularly in the carbon domain, is nowhere near as good as that in a 1-D spectrum. You might remember that we recommended a typical data matrix size of 2k (proton) × 256 (carbon). There are two persuasive reasons for limiting the size of the data matrix you acquire – the time taken to acquire it and the sheer size of the thing when you have acquired it. This data is generally artificially enhanced by linear prediction and zero-filling, but even so, this is at best equivalent to 2k in the carbon domain. This is in stark contrast to the 32 or even 64k of data points that a 1-D ^{13}C would typically be acquired into. For this reason, it is quite possible to encounter molecules with carbons that have very close chemical shifts which do not resolve in the 2-D spectra but will resolve in the 1-D spectrum. So the 1-D experiment still has its place…

9.4   Piecing It All Together

As we've mentioned before, the interpretation of NMR spectra is often made complex by the sheer quantity of information that you are confronted with. This is every bit as true for carbon NMR as it is for proton and when you combine the two, that huge pile of information just gets bigger… More important still then that you approach the pile in a logical, methodical manner.

Once your problem takes on a ^{13}C dimension, you are of course, obligated to examine the ^{13}C data with the same level of dispassionate scrutiny that you apply to the proton data. Chemical shifts cannot be fudged and unexpected peaks cannot be glossed over. You have to be able to account for everything you see to have confidence in your product.

We will assume that you have already been through the proton data with a fine-toothed comb and found it wanting in some way, or insufficient to give the level of reassurance that you require. So turning to carbon, a 1-D ^{13}C spectrum of adequate signal-to-noise would be a luxury, though not an absolute necessity. We'll assume you have one. Strike out the carbons for any known solvents, etc. and then count the total number of carbon peaks in the spectrum. Do they match the requirements for your proposed structure? (Don't forget that a para-di-substituted aromatic ring gives four peaks for its six carbons on account of its symmetry.) Note also that knowing with certainty, the number of carbons in a structure, can be very helpful in narrowing the search for a molecular formula by mass spectroscopy (accurate mass).

If you have a DEPT 135 spectrum, now is the time to use it. Categorise all peaks to one of the following types: quaternaries and carbonyls, methines, methylenes and methyls. Now get hold of plausible prediction data for the shifts of your proposed structure. Use HMQC/HSQC spectra to assign the proton-bearing carbons and if satisfactory, move on to assign all the quaternaries and carbonyls by using the HMBC spectrum. Do all the long-range connectivities from the HMBC make sense? Does it all hang together?

As with proton interpretation, this must be considered an iterative process. Try to shoot your proposed structure down. Don't be afraid to tear it up at any stage and start again if some glaring problem becomes apparent. Resist temptation – don't hammer the square peg into a round hole! This is why we do spectroscopy in the first place. If it crashes and burns then it was wrong so shed no tears. If it survives, then it's got a good chance of being a winner. Finally, go back again and check that there is no mismatch between any carbon data and any supplementary proton data, e.g. NOE experiments.

When it **all** sits happily or can at least be explained, the job is done as well as it can be. Not before.

9.5 Choosing the Right Tool

If you have successfully read this far, it might have occurred to you that some problems could well be solved by either an NOE-based approach, or by an HMBC approach and you might be wondering which technique would be preferable under such circumstances. In truth, there may not be a right or wrong answer to this question and each problem should be considered on its merits. The selection of experiment may even be down to personal preference or to the hardware available to you. Questions of positional isomerism can often be resolved by either approach. We have seen how our naphthalene problem could be resolved by using an NOE technique.

This problem could also have been resolved by an HMBC approach – you would expect to see a correlation from the protons of either the $-CH_2-$, or the $-CH_3$ to one of the quaternary carbons at the junction of the two rings. This same carbon should also show correlations to at least two, and ideally three of the protons on the un-substituted aromatic ring and one of the protons on the substituted ring.

It is when questions of stereochemistry arise that the NOE techniques come into their own. For example, consider the compounds below. There would be no chance of resolving these two structures by HMBC, but an NOE technique might well prove successful. (The methyl group would be expected to give an enhancement to either of the –CH$_2$–OH protons in one isomer or to the >CH–CH$_2$OH in the other, depending on which face of the ring the two substituents lie relative to each other.)

As in the case of all NMR problem-solving, the issue is always one of using the most appropriate tool for the job. The two techniques are in no way mutually exclusive. Too much data is not a bad thing if the instrument time is available but taking a chance on insufficient data can be a costly mistake in the long run...

Tables 9.1 through to 9.7 give a useful guide to ^{13}C chemical shifts...

Table 9.1 ^{13}C chemical shifts of some common carbonyl functions.

Type of carbonyl	Typical shift	Type of carbonyl	Typical shift
	205–210		170–180
	195–200		165–170
	~195		170–175
	196–202		~168
	190–195		160–165
	167–173		160–165
	165–172		153–160

Note: Thio-carbonyl analogues generally absorb at considerably lower field – sometimes by as much as 40 ppm.

Table 9.2 ^{13}C chemical shifts of some CN functions.

Type of CN function	Typical shift
R—≡N / ⬡—≡N	115–120
R—N$^+$≡C$^-$ / ⬡—N$^+$≡C$^-$	156–166
R—S—≡N / ⬡—S—≡N	110–115
R—CH=N—R' / ⬡—CH=N—R	155–170

Table 9.3 Data for the estimation of aryl ^{13}C chemical shifts.

Substituent X	C1	C2	C3	C4
–H	0.0	0.0	0.0	0.0
–CH$_3$	9.2	0.7	0	–3.0
–CH$_2$-any (approx.)	2–12	–2 to (+)2	–2 to (+)2	–2 to (+)2
–CH=CH$_2$	9	–2	0	–0.8
–CΞC–R (approx.)	–6	4	0	0
–Phenyl	8	–1	0.5	–1
–F	34	–13	1.5	–4
–Cl	5	0	1	–2
–Br	–5	3	2	–1
–I	–31	9	2	–1
–OH/–OR (approx.)	30	–13	1	–7.5
–O–Phenyl	28	–11	0	–7
–OCOCH$_3$	22	–7	0	–3
–NH$_2$/–NR$_2$ (approx.)	17	–14	1	–10
–NH$_3$$^+NR_2H^+$ (approx.)	3	–5	2	1
–NO$_2$	20	–5	1	6
–CN	–16	3.5	1	4
–NC	–2	–2	1	1
–SH/–SR (approx.)	7	0	0	–3.5
–S–Phenyl	7	2.5	0.5	–1.5
-SOR	18	–5	1	2
-SO$_2$R	12	–1	1	5
-SO$_2$Cl	16	–2	1	7
-SO$_3$H	15	–2	1	4
-SO$_2$NH$_2$	11	–3	0	3
-CHO	8	1	0.5	6
-COR	9	0	0	4
-COOH/-COOR (approx.)	2	1.5	0	4.5
-CONH$_2$/-CONR$_2$ (approx.)	5.5	–1	0	2

Note: Substitute values relative to benzene (128 ppm) as follows: Chemical shift of C1–4 = 128 + additive value for C1–4 from table above.

Table 9.4   ¹³C chemical shifts of some common heterocyclic and fused aryl compounds.

Table 9.5   Data for the estimation of alkene ¹³C chemical shifts.

Substituent X	C1	C2	Substituent X	C1	C2
–H	0	0	–OCOCH₃	18	−27
–alkyl	10–20	−4 to −12	–NR₂	28	−32
–CH=CH₂	14	−7	–N⁺R₃	20	−11
–CH≡CH	−6	6	–NO₂	22	−1
–Phenyl	12.5	−11	–CN	−15	14
–F	25	−34	–NC	−4	−3
–Cl	3	−6	–SR	9	−13
–Br	−9	−1	–CHO	15	15
–I	−38	7	–COR	14	5
–OR	28	−37	–COOR	5	10

Note: Substitute values relative to ethene (123 ppm) as follows: Chemical shift of C1 and C2 = 123 + additive value for C1/C2 from table above.

Table 9.6 ^{13}C chemical shifts for alkynes.

Type of alkyne	Typical shift
R———≡———R'	75–80
	~85 (C1) ~80 (C2)
	~90

Table 9.7 Data for the estimation of alkyl ^{13}C chemical shifts.

Substituent X	α	β	γ
–H	0	0	0
–Alkyl	9	9	–3
–C=C–R$_2$	20	7	–2
–C≡C–R	4	6	–3
–Phenyl	22	9	–3
–F	70	8	–7
–Cl	31	10	–5
–Br	19	11	–4
–I	–7	11	–2
–OR	49	10	–6
–OCOR	57	7	–6
–NR$_2$	28	11	–5
–NR$^+_3$	28	6	–6
–NO$_2$	62	3	–5
–CN	3	2	–3
–SR	11	11	–4
–SO$_2$R	30	7	–4
–CHO	30	–1	–3
–COR	23	3	–3
–COOR	20	2	–3
–CONR$_2$	22	3	–3
–COCl	33	8	–3

Note: This table gives only very approximate shift estimates and is intended for use as a rough guide only. The presence of highly branched substituents and atoms bearing multiple halogens, multiple oxygen atoms etc. can cause even more serious deviations rendering the table of questionable value under such circumstances. It is used by summing the substituent effects at each carbon relative to methane (–2.3 ppm). For example, shift of carbon 'a' below would be estimated as: –2.3 + 49 + 9 + 9 = 65 ppm (approx.). Actual value 63 ppm.

10

Nitrogen-15 NMR Spectroscopy

10.1 Introduction

In the first edition of this book, ^{15}N spectroscopy was dealt with in a chapter entitled, 'Some of the Other Nuclei'. On reflection, this hardly did the nucleus justice and as the technique has become increasingly important as a problem-solving tool over the years, we thought the nucleus well worthy of a chapter all to itself.

Before going any further, it's worth noting that every nucleus that can be studied by NMR has its own particular properties which have to be considered when studying that nucleus – and ^{15}N is certainly no exception. For a start, the natural abundance of ^{15}N is extremely low at only 0.36%. This, combined with a particularly low frequency (and therefore, energy) signal, means that the overall sensitivity of the ^{15}N signal is only about 1/10 that of the ^{13}C signal. For this reason, direct observation of the ^{15}N nucleus is not a practical proposition (unless of course, you have a ^{15}N enriched compound to work with!), and indirect detection in a ^{15}N HMBC experiment is the only way to go. Of course, all the early work on the ^{15}N nucleus would have been carried out on enriched samples but if you are working with some unknown reaction product, re-synthesising a ^{15}N enriched version of it would be unlikely to offer a viable way forward. It's the ability to acquire natural abundance ^{15}N NMR that now makes it a really useful problem-solving tool.

The development of high-field magnets, cryoprobes with inverse detection, and non-uniform sampling (dealt with in Chapter 3) has made this possible and brought the acquisition of ^{15}N NMR data out of the 'might get something if we leave it scanning all weekend' realm and into that of the 'routine'. High-quality ^{15}N HMBC data can now be obtained in about an hour on a 600 MHz system with a cryoprobe, for samples of about 5–10 mg. The chapter is a long one and contains fairly comprehensive information on all the main nitrogen groups and on quite a few less common ones as well. This is partly because definitive published ^{15}N data isn't that easy to come by and because we have found it an interesting learning experience in our own area of work.

In addition to the extremely low sensitivity of the nucleus, another problem that it presents us with is a very large range of frequencies over which nitrogens in different environments can resonate. While alkyl amines, for example, resonate at one end of the spectrum, nitroso compounds resonate at the other – and there can be almost a thousand ppm between them! This huge spectral width does present problems. If we were to attempt to collect data over a full range of 1000 ppm in the ^{15}N domain, we would end up with enormous and unwieldy data sets, most of which wouldn't contain anything at all. Such data sets would take longer to acquire and require excessive storage space. Furthermore, hardware limitations prevent the uniform irradiation of such a wide spectral range, so for these reasons, it is advisable to acquire over a range of about 0–400 ppm in the nitrogen domain. Should you encounter a nitrogen outside this range, it won't just vanish from your spectrum as it will be 'folded' in the nitrogen dimension and it will be necessary to re-acquire the data with a different frequency offset to confirm this and observe its true chemical shift.

10.2   Referencing

As far as a 'standard' for the ^{15}N nucleus is concerned, the situation is not as clear cut as it is for proton and carbon NMR, where TMS is universally accepted. There are two standards in use in the ^{15}N arena and both have certain advantages and disadvantages. As with standards used for any other nuclei, we want a signal that is out of the way of all, or at least most, of the typical resonances encountered, so this means it will need to be at one end of the spectrum or the other. The two standards used in ^{15}N NMR are liquid ammonia and nitromethane.

The advantage of liquid ammonia is that it gives a signal above all other commonly encountered groups and therefore, since its position is assigned to 0 ppm, all other resonances will be downfield relative to it and therefore will all have positive chemical shift frequencies. The disadvantage of liquid ammonia is of course that it is difficult to handle and difficult to dispense into an NMR tube at the right concentration that can be observed – without swamping the spectrum. More difficult than you might think…

Nitromethane is much easier to handle and that is undoubtedly an advantage – but setting a low-field signal to 0 ppm means that all commonly encountered resonances will be negative. Higher field signals will have increasingly large negative chemical shifts. This is somewhat counter-intuitive and will for many, cause confusion. And there is worse – as we'll see in the following pages, some N-bearing moieties (the nitroso, N-nitroso and certain N-bearing heterocycles spring to mind) have resonances at even lower field than nitromethane. This could lead to spectra containing both positive and negative chemical shifts!

All chemical shifts quoted in this book are quoted on the liquid ammonia scale. We believe that the best solution to these problems is to use the liquid ammonia scale but without actually attempting to add liquid ammonia to the sample. Default referencing calculated from the frequency offset of the instrument is accurate enough for real-world problem solving.

10.3   Using ^{15}N Data

The comparatively recent ease of data acquisition has opened up great possibilities for problem solving in cases where other techniques might be found wanting. Whereas the use of ^{15}N HMBC to establish connectivities within a compound is straightforward enough, there is the potential to use the technique in a more sophisticated manner. There is a rich source of information to be gathered in terms of the subtleties of ^{15}N chemical shifts and the intensities of the correlations that give rise to them and we will be exploring these ideas in detail in this chapter. These parameters can provide a further level of confidence in the assignment of a structure and can indeed be the difference between a definitive solution to a problem and no solution at all. For example, the two isomers shown below were differentiated by using ^{15}N HMBC where prospects for success using other methods looked bleak.

With the amine function ortho- to the pyridine nitrogen, a correlation between H18 and N12 was observed showing it to have a shift of 256 ppm. Where the amine is para- to the pyridine nitrogen, its shielding influence is diminished, N1 showing a shift of 269 ppm (by correlation from H3). This shift difference might be relatively small but it is nonetheless significant enough to allow the two regioisomers to be differentiated with confidence.

The chemical shifts of the amine nitrogens also provide useful complementary information which gives an extra level of confidence in the solution. In the case of the ortho-substituted amine, the proximity of the pyridine nitrogen exerts a greater deshielding influence (on the amine nitrogen) than in the para-substituted amine. (ortho amine, N18, 107 ppm, para amine, N8, 100 ppm). Again, the shift difference is small but significant.

In order to exploit the possibilities afforded by ^{15}N HMBC, it is necessary to develop an understanding of the factors that influence shifts and to establish reliable trends that can be used in related compounds. This chapter will hopefully prove useful as it attempts to establish such trends. (Note that as in the case of ^{13}C spectroscopy, and for the same reasons, ^{15}N shifts are driven very much by electron distribution and molecular anisotropy is of little significance which makes the task of shift prediction considerably safer)

It's worth noting that the 'weapon of choice' for the acquisition of ^{15}N data is the ^{15}N HMBC experiment, preferably with inverse detection as has been mentioned earlier, optimised for 7 Hz couplings. This value gives the best compromise for the detection of 2- and 3-bond correlations in the majority of commonly encountered compounds. It is perfectly possible to run ^{15}N HSQC spectra but these are of little value for a number of reasons. Firstly, not all N–H protons will show correlations to the nitrogens they are bound to, since they are often severely broadened by chemical exchange, so a ^{15}N HSQC may end up containing relatively little useful information. Secondly, the ^{15}N HMBC pulse sequences used do not suppress the 1-bond correlations that are often readily observed in these spectra. This is quite deliberate as 1-bond correlations will seldom cause confusion since there are generally not that many signals in a ^{15}N spectrum and they can be easily identified for what they are. So the ^{15}N HMBC will usually contain both long- and short-range correlations all in the one experiment. Excellent! (This is in marked contrast to the ^{13}C HMBC experiment where 1-bond 'break-through' can cause problems, particularly in the aromatic region which can often be extremely 'busy' with many carbons showing correlations in the 7–8/110–140 ppm region. 1-bond correlations have to be identified and 'edited out' of your spectrum before assignment can begin. Since these '1-bonders' can often align with possible positions of desired 2- and 3-bond correlations, the potential for confusion will be self-evident!)

The data used to establish both chemical shift predictions and the trends within them comes largely from our own observations and has been checked against literature and ACD predictions (using the ACD XNMR DB). Since our aim has been to provide guidance in terms of chemical shift expectation, greater weighting has been given to data acquired in our preferred solvents, $CDCl_3$ and D_6 DMSO, in preference to examples acquired in D_6 acetone, D_6 benzene, D_5 pyridine, D_4 methanol or neat liquids.

Spectrum 10.1 which was recorded on a 600 MHz instrument fitted with a cryoprobe, shows an example of a ^{15}N HMBC. The compound in question is of the type shown below:

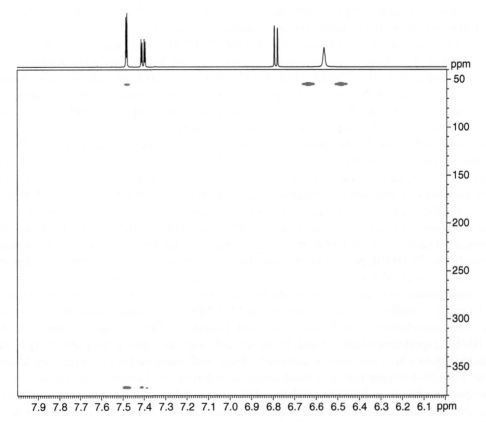

The chemical shift of the amine nitrogen is 55 ppm and shows a clear 3-bond correlation to the aromatic proton giving a fine doublet at 7.49 ppm. There is also a strong, and in this case, very useful, 1-bond correlation to this nitrogen from the amine proton itself. Note that whether or not

Spectrum 10.1 ^{15}N HMBC.

you see 1-bond correlations depends largely on how broad the –NH signal is in the proton domain. The sharper the –NHs, the more likely you are to see them. As with ^{13}C HMBC, 2-bond correlations can sometimes be quite weak and that is so in this case as there is no obvious correlation to be seen from the methylene protons adjacent to the amine.

The nitrogen of the nitro group has a signal at 371 ppm and shows 3-bond correlations from both the aromatic protons flanking the group (7.49 and 7.41 ppm). The common correlation from the former signal to both nitrogens confirms the regiochemistry of the structure.

Please note the following...
Nitrogen-bearing groups are listed alphabetically with the exception of amines which are listed before amides in order to establish simple nitrogen behaviour before the consideration of more complex environments. A number of N-bearing compounds are listed in Section 10.27 which aren't easily described. These feature nitrogens in more unusual and exotic environments.

All chemical shifts are quoted relative to liquid ammonia (0 ppm).

Shifts are described as 'approx.' to account for variations in solvents used and concentration of samples as well as possible referencing inaccuracies and the limitations of available, relevant analogues.

The term 'downfield' is used to describe a shift to larger ppm number (and vice versa) and 'lower than' or 'to lower field' describes a movement in this direction.

In cases where correlations are described as 'strong' or 'weak', these terms are used in the context of the ^{15}N HMBC experiment optimised for 7 Hz ^{1}H–^{15}N couplings.

At the end of some of the sections that follow, we have included *'General observations'*. These are relevant to that particular group or type of compound and though they may be anecdotal to some extent, they are nonetheless potentially useful.

10.4   Amines

10.4.1   Alkyl

See also Section 10.23.

Primary amines (alkyl) N1 expected at approx. **24** ppm. In the case of protonation, relatively small but significant downfield shift noted. Expected value approx. **31** ppm. (Note the common impurity, ammonium chloride where the NH_4 + ion gives a shift of approx. 25 ppm.)

Secondary amines (alkyl) N1 expected at approx. **38** ppm. Protonation expected to bring value of N1 down to approx. **46** ppm.

Tertiary amines (alkyl) N1 expected at approx. **46** ppm. Protonation expected to bring value of N1 down to approx. **54** ppm.

Quaternary amines (alkyl) N1 expected at around 65 ppm.

General observations: 2- and 3-bond correlations expected but correlations can be weak as correlating proton signals are often wide multiplets. Chemical shifts tend to increase with increasing alkylation of the nitrogen (as a rough guide, approx. 10 ppm per group). Protonation also increases chemical shift to a small but significant extent (again, approx. 10 ppm) and this could be used to estimate the degree of protonation if no other method were available. Note also that the most upfield of amines can be 'folded' as they may well be outside the normal sweep width in the nitrogen domain on your spectrometer.

   Cyclic amines In the case of cyclic amines, chemical shifts are broadly in line with their open chain counterparts (free bases and salts), though small deviations can be expected due to individual anisotropies.

N1 approx. 38 ppm.   N1 approx. 47 ppm.   N1 approx. 53 ppm.

In the special case of aziridine, shown below, the peculiar anisotropies and extreme bond strain, contribute to an extreme chemical shift of approx. −10 ppm.

Alkyl substitution on the aziridine nitrogen gives rise to trends similar to those already noted. In the four-membered analogue, a smaller but notable upfield shift is observed giving a shift of approx. 25 ppm, but this trend extends no further – the five-membered analogue gives a shift indistinguishable from the open-chain compound.

General observations: It should be noted that –CH_2–R alkyl groups are significantly more deshielding than –CH_3 groups and that branched chain hydrocarbons attached to the nitrogen at the branch point are more deshielding still e.g. the piperidine compounds noted earlier. These observations are valid for open-chain and cyclic amines.

10.4.2   Aryl

Primary amines (aryl) N7 expected at around 60 ppm. Protonation tends to move N7 upfield, in this case to approx. 50 ppm.

General observations: Substitution with electron-donating substituents (e.g. –OR) in the ortho or para positions, moves N7 upfield by up to 10 ppm (i.e. approx. 50 ppm), though meta substitution with an electron-donating group has little effect.

Substitution with electron withdrawing groups (e.g. –COR or –COOR) ortho or para to N7 has a downfield effect on its shift of approx. 10 ppm giving rise to shifts of approx. 70 ppm. Once again, meta substitution with such groups has little influence on N7.

It is interesting to note the upfield shift of aryl amines on protonation which is in marked contrast to the alkyl amines.

Secondary amines (aryl) N7 shows a small upfield shift on methylation to approx. 55 ppm, again reversing the trend seen in alky amines. Alkylation with a longer chain alkyl group moves N7 to a lower field (~66 ppm) as noted in other amines. Branched chain hydrocarbons are more deshielding still, as noted previously. Protonation appears to have little influence on shift.

Tertiary amines (aryl) N7 shows a significant upfield shift on di-methylation to approx. 45 ppm, though data suggests that di-alkylation with longer alkyl groups brings about significant depression of the nitrogen shift (~73 ppm). Protonation (of the N-di-methyl aniline) moves N7 downfield to approx. 52 ppm.

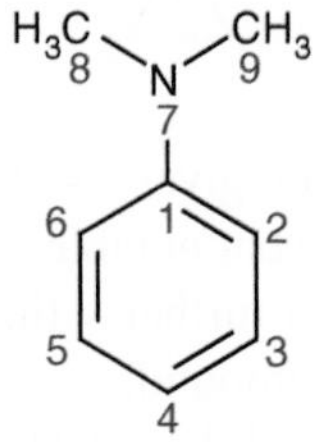

Quaternary amines (aryl) N7 shows a shift of approx. 58 ppm.

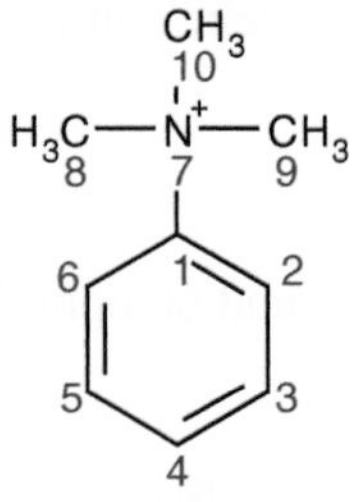

General observations: Confirming chemical shift trends for secondary, tertiary and quaternary aryl amines is more difficult as a general pattern is less clear. Longer chain alkyl groups (i.e. C2 and above) lead to significant deshielding of amine nitrogen shifts, in comparison with methylated species.

Secondary amines (bi-aryl) N7 shows a shift of approx. 91 ppm.

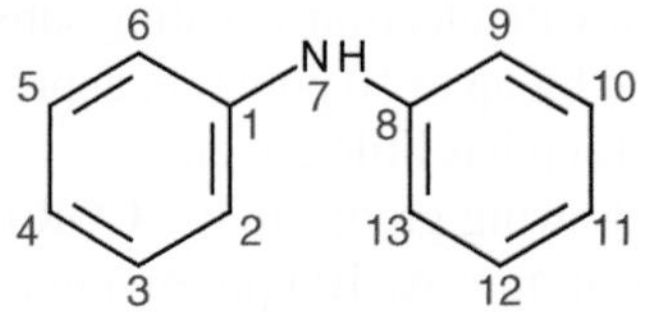

In the event of bridging between the rings, nitrogen chemical shift is shifted downfield as shown below.

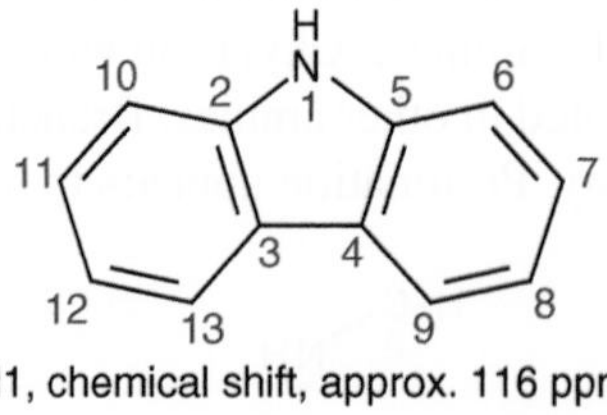

N1, chemical shift, approx. 116 ppm.

General observations: The second aryl ring causes significant depression of chemical shift. There is insufficient data available to comment on either the influence of protonation or further aryl substitution of the nitrogen.

10.5 Conjugated Amines

Conjugation brings about a considerable downfield shift in the ^{15}N shift of amines.

N1 would have a predicted shift of approx. 80 ppm with little variation on
changing alkene stereochemistry.

General observations: While there is a lack of data to draw definitive conclusions, it would seem reasonable to conclude that other influences (protonation, further alkylation, etc.) might act in an approximately additive manner, enabling at least approximate estimates of shifts to be made.

10.6 Amides

Primary amides (alkyl) Chemical shift of N3, approx. 116 ppm. There is no significant change noted in the case of extended alkyl chains.

Secondary amides (alkyl) Chemical shift of nitrogen moves upfield to approx. 104 ppm on alkylation of nitrogen by a methyl group but moves downfield slightly where the alkyl group is an alkyl chain e.g. n-propyl analogue gives a shift of approx. 120 ppm. A branched hydrocarbon e.g. isopropyl causes a further increase to the ^{15}N shift (136 ppm).

Tertiary amides (alkyl) Chemical shift of nitrogen moves upfield to approx. 98 ppm in the case of alkylation with two methyl groups. Significant depression of chemical shift in the event of N-substitution with alkyl chains (approx. 125 ppm) is observed.

General observations: NH$_2$/NH signals, while often slightly broadened in the proton domain, can often give strong 1-bond correlations in ^{15}N HMBC experiments which can be very useful as a confirmatory assignment tool. It is possible, in the case of tertiary amides where the N-substituents are not the same, to observe slightly different shifts for the nitrogen in the two different rotameric forms. This difference is likely to be small, possibly less than 2 ppm. (See Section 10.28.)

Primary amides (on aryl rings) The chemical shift of N9 (if observed by 1-bond correlation) will be at approx. 102 ppm. There is little variation to be expected shift with aryl substitution.

Secondary and tertiary alkyl amides (on aryl rings) Where the alkylating group is the methyl group, very little impact is noted on the shift of N9 (as above). Where the alkylation group(s) are alkyl chains, a downfield shift to about 122 ppm can be expected.

Secondary and tertiary aryl amides (on aryl rings) Where the N-substituent is an aryl ring, the chemical shift of the amide nitrogen will be approx. 133 ppm, though data available is not sufficient to draw firm conclusions for bi-aryl tertiary amides.

Conjugated amides (alkyl and aryl) Conjugating chains on the carbonyl side of amides (secondary and tertiary, alkyl and aryl) have little or no discernible influence on ^{15}N shifts. On the nitrogen side, however, conjugating chains can be expected to contribute an approx. 20–25 ppm downfield shift to amide nitrogens i.e. 140–145 ppm.

General observations: The impact of substituents on the carbonyl group is largely insignificant while substituents on the nitrogen have considerable influence on the ^{15}N shifts of amides. Alkyl chains deshield to a greater extent than methyl groups and conjugating chains further deshield amide nitrogens.

10.7 Amidines

An example is shown below. (R = alkyl group)

The compound was considered to be a salt (the amidine group is highly basic). N10 and N12 presented as a single absorbance at 104 ppm. The correlations were of the combined 1- and 3-bond type.

General observations: As amidine nitrogens are inter-related by tautomerism, whether they are observed individually, or as a combined signal will depend on the rate of the tautomeric interconversion on the NMR timescale. In the example above, tautomeric interconversion was fast on the NMR timescale, probably enhanced by the presence of an acid, as only a single absorbance was observed. Prediction offers distinct chemical shifts for each of the amidine nitrogens (in the example above, approx. 63 ppm for N10 and approx. 200 ppm for N12). The experimental result of 104 ppm represents a time-averaged value for both environments.

10.8 Azides

The chemical shifts of all nitrogens in an azide group are seldom observable for obvious reasons but the example below (acquired for several hours on a strong sample), revealed correlations to all three nitrogens.

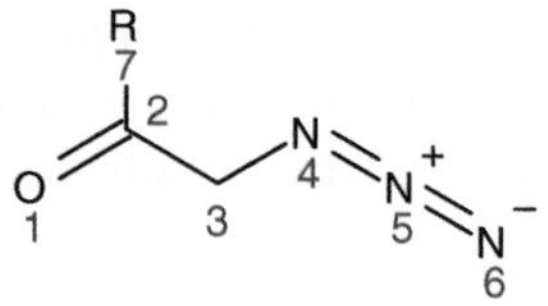

In a compound of this type (R = alkyl), H3 showed correlations to N4 (65 ppm), N5 (247 ppm) and a 4-bond correlation to N6 (208 ppm). In another example, shown below, N6 was shown to correlate to H2 and H4 with a shift of 82 ppm, suggesting that this nitrogen is quite sensitive to the group it's attached to. (R = alkyl)

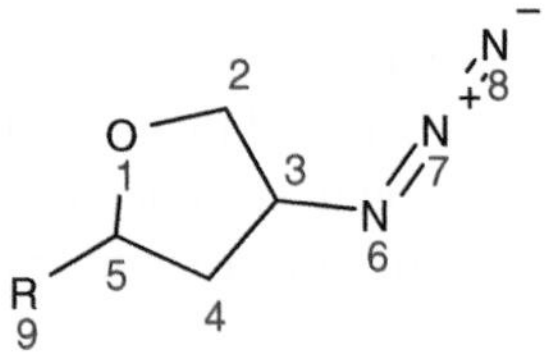

10.9 Carbamates

Primary carbamates show ^{15}N shifts typically in the region of 74 ppm. Methylation of the nitrogen (secondary carbamates) gives rise to a very slight upfield shift (approx. 71 ppm) while an alkyl chain brings about an approx. 10 ppm downfield shift (approx. 84 ppm). The secondary carbamate of the type shown below gives a clear indication that a branched chain hydrocarbon (R) attached to the nitrogen at the branch point, adds considerably to the chemical shift. A 1-bond correlation from H7 was noted (N7 99 ppm).

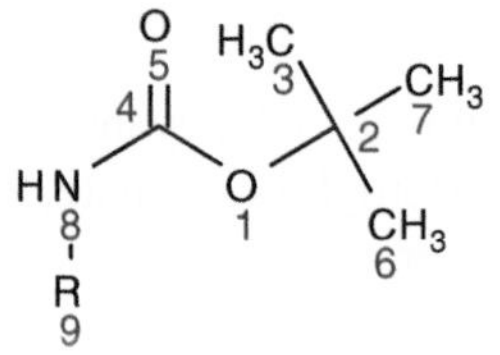

Di-methylation (tertiary carbamates) brings about a further upfield shift to approx. 65 ppm in the compound shown below.

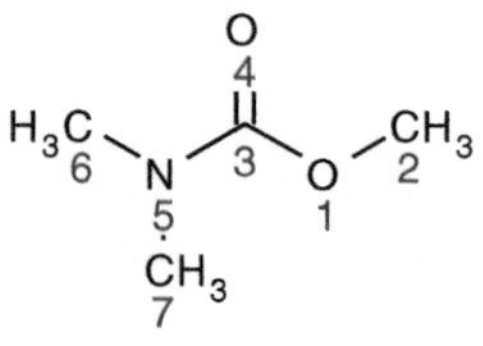

A similar tertiary carbamate, featuring two isopropyl groups on the nitrogen has been recorded at 112 ppm once again underpinning the influence of branched hydrocarbons.

Conjugation on the oxygen side of the carbamate would appear to have little influence on the carbamate nitrogen shift though data is sparce. Conjugation on the nitrogen is more difficult to assess as the tautomeric –C=N– tautomer may well predominate.

In the case of aryl (secondary) carbamates, a significant downfield shift can be expected. For example, the carbamate below has a nitrogen shift of approx. 104 ppm, while the addition of a further aryl ring (to give a tertiary bi-aryl carbamate) would give a shift of approx. 120 ppm. This figure would be much the same for a tertiary carbamate where the nitrogen is bonded to one aryl ring and one alkyl chain.

General observations: In general, carbamates appear to mirror amide behaviour but their chemical shifts tend to be approx. 25 ppm higher field than the amide counterpart. As with amides, the NH proton signals are often sharp enough (in suitable solvents) to provide useful 1-bond correlations which can be helpful for confirming assignments.

10.10   Cyanates and Thiocyanates

Methyl cyanate has a reported shift of 158 ppm with virtually no change noted in the ethyl analogue which is not surprising in view of the distance between the alkyl group and the nitrogen. The phenyl analogue shows only a slight downfield shift with a reported value of approx. 170 ppm.

In the case of the related sulfur compounds (thiocyanates), the sulfur depresses the nitrogen shift by about 110 ppm in each case. (Methyl thiocyanate, 275 ppm, ethyl thiocyanate and phenyl thiocyanate both recorded at 280 ppm.)

(See also, isocyanates and isothiocyanates.)

General observations: The recording of nitrogen data for these compounds would require a modified HMBC pulse sequence as in each case, the nitrogen will be more than three bonds away from the closest proton.

10.11 Diazo Compounds

Diazo nitrogens are characterised by their extremely high chemical shifts. The di-ethyl compound below has a shift of approx. 535 ppm.

The bi-aryl analogue has a chemical shift of 510 ppm, while in the unusual cyclic compound shown below, N1/2 showed a chemical shift of 464 ppm (R = alkyl chain).

General observations: Despite the limited data available, it seems that aryl-substitution of diazo nitrogens leads to upfield shifts of approx. 25 ppm, relative to purely alkyl substituted species. It should be noted that the ACD database does not offer clear distinction between cis and trans stereochemistry of these compounds. In highly strained cyclic systems, a large upfield shift can be expected and conjugation of the diazo function gives rise to an even larger upfield shift of as much as 100 ppm though data limitations prevent a full investigation of this. All diazo nitrogen shifts will be 'folded' if using typical acquisition parameters.

10.12 Formamides

The most familiar formamide is undoubtedly the solvent, dimethyl formamide (DMF), which has a well-recognised ^{15}N shift in the region of 105 ppm.

In other formamides, the restricted rotation of the carbonyl–nitrogen bond (N1–C10 below) gives rise to rotamers which are distinct on the NMR timescale. The differing anisotropies of these distinct rotameric species give chemical shifts of 144 and 148 ppm for N1. In both rotamers, correlations were observed from H3 and 10a. (See Section 10.28 for further discussion of restricted rotation.)

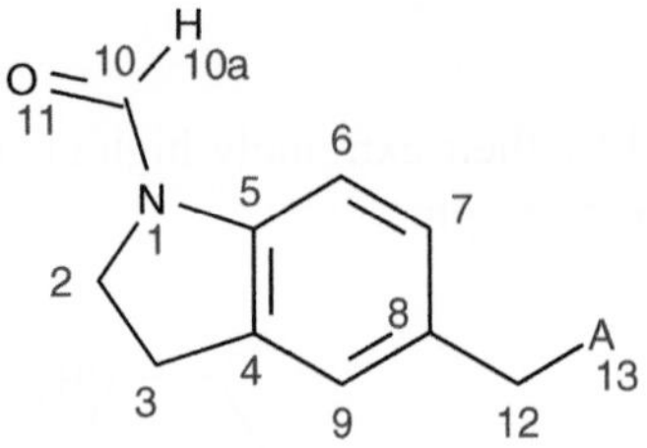

A = any group, not specified.

In a related compound (see below), H-bonding (between H11a and N7) has been seen to stabilise just a single rotamer. [This H-bond gave rise to an interesting long-range correlation between H11a and N7 (271 ppm)]. This phenomenon will also be covered in detail in Section 10.28. H3 and H11a showed correlations to N1 (144 ppm).

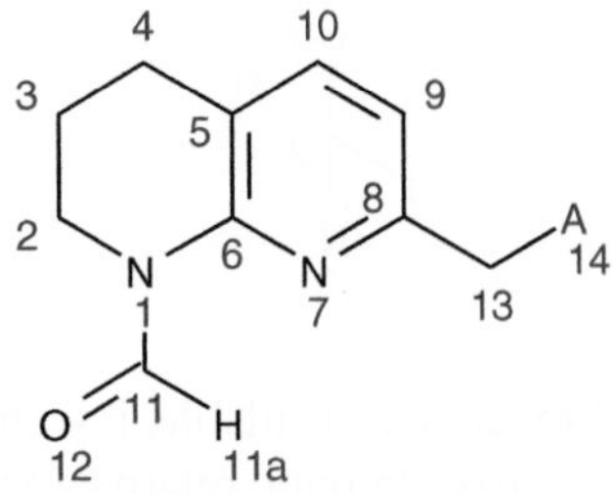

A = any group, not specified.

In the simple aryl formamide shown below, N7 has a chemical shift of approx. 141 ppm.

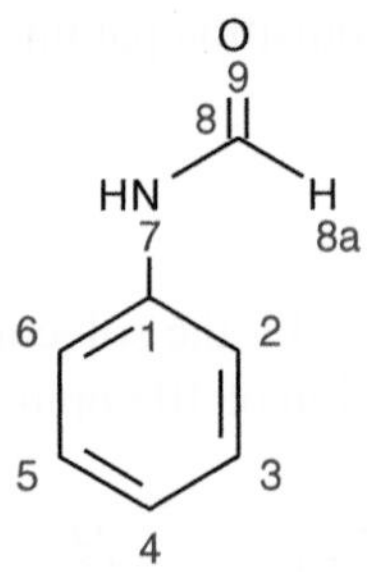

General observations: There is a general lack of data on alkyl formamides related to DMF but it would seem reasonable to conclude that alkyl chains in place of methyl groups would be likely to increase the chemical shift of the formamide nitrogen, and hydrocarbons branched at the point of attachment to the nitrogen, to increase it further.

10.13   Hydrazines

In the case of alkyl hydrazines, shifts of approx. 50 ppm for N1 and 75 ppm for N2 might be expected, though terminal hydrazines are not well represented in literature.

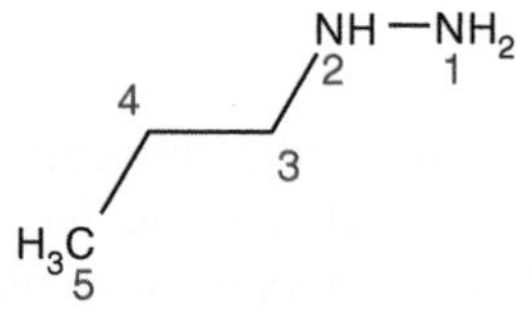

Secondary alkylation (at N1) would be expected to bring N1 to a similar value (approx. 75 ppm).

In the case of the aryl hydrazine (below), chemical shifts of 85 and 60 ppm can be expected for N7 and N8 respectively, while the symmetrical bi-arylated compound has a shift of approx. 96 ppm.

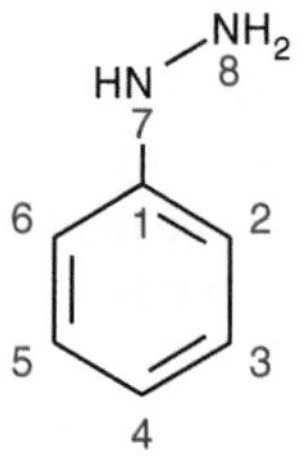

General observations: Though reliable examples are relatively scarce, it would seem likely that further alkylation would yield nitrogen shifts in-keeping with the trends seen in other classes of compounds. That is, chemical shift showing the greatest increase when alkylated with a conjugated chain, followed by branched hydrocarbon, ordinary alkyl chain, with the methyl group being the least deshielding.

10.14   Hydroxamic Acids

The hydroxamic acid nitrogen in a compound of the type shown below was recorded at 166 ppm(correlated to H5).

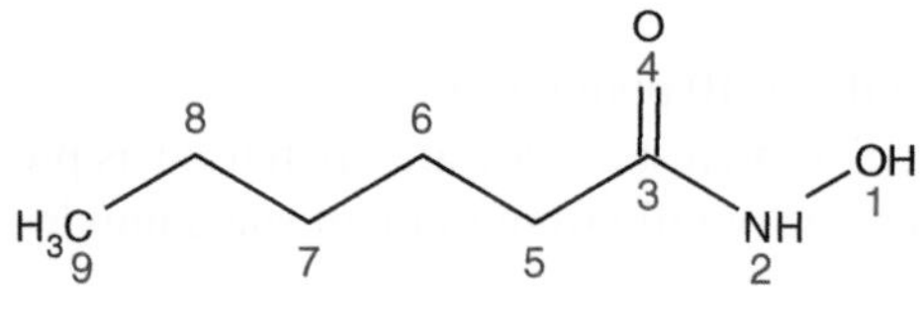

The aryl hydroxamic acid shown below, showed a near identical shift of 165 ppm.

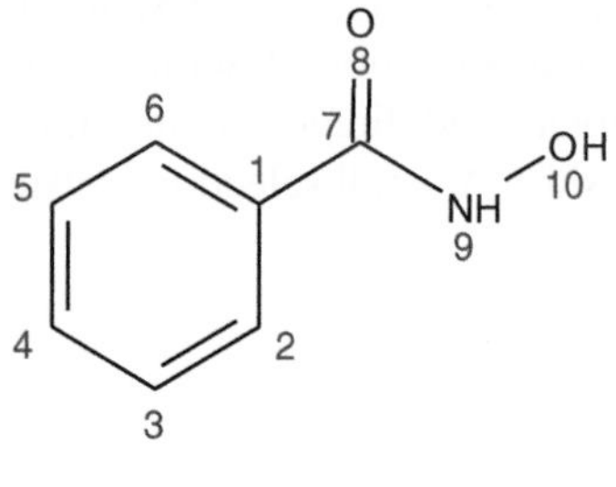

10.15   Hydroxylamines

The hydroxylamine nitrogen in the methylated compound shown below has been reported to have a shift of 128 ppm, while the isopropyl analogue has been reported at 146 ppm and the tertiary butyl analogue at approx. 148 ppm. A value of around 135 ppm would be highly likely for straight chain alkyl groups given the trends observed with other functionalities.

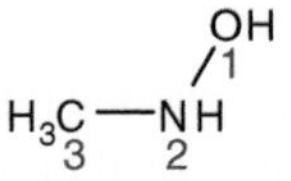

10.16   Imides (Alkyl and Aryl)

N3 of the imide shown below would have a chemical shift of approx. 165 ppm. Methylation of the nitrogen would be unlikely to bring about any significant change to the shift while alkylation with an alkyl chain (C2 or above) would be likely to depress the nitrogen shift by up to 20 ppm, though no definitive data is available on this.

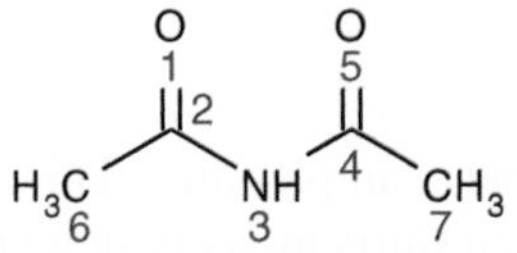

Aryl substitution depresses the shift of N3 to approx. 175 ppm.

A similar shift can be observed in the simple cyclic imide (succinimide) shown below (176 ppm).

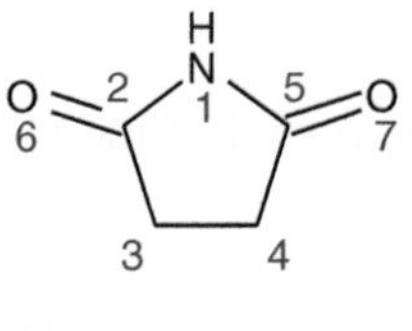

Arylation of N1 will yield shifts 5–10 ppm lower.

Phthalimides are of particular interest as they are often used as protecting groups. Chemical shifts seen in these compounds vary little from those of succinimides discussed above.

10.17   Imines

In common with other >C=N– species, imines give rise to highly deshielded nitrogens with chemical shifts normally observed at >300 ppm. Useful examples are limited but substitution both on the carbon side of the moiety and directly on the imine nitrogen itself appear to have limited influence on the nitrogen shift. A small selection is listed here...

N7, 325ppm. (R = alkyl)

N7, 324 ppm.

N8, 308 ppm.

Even C-substitution with a conjugated species has little impact on the imine nitrogen shift…

N7, 310 ppm.

10.18 Isocyanates and Isothiocyanates

In the case of methyl isocyanate (below), a shift of approx. 19 ppm has been recorded while the ethyl analogue gives a shift of approx. 35 ppm and the phenyl compound, a shift of approx. 50 ppm.

The corresponding isothiocyanates give values of approx. 93 ppm for the methyl compound, and 110 ppm for both the ethyl and phenyl compounds.

General observations: The nitrogen shifts observed show some sensitivity to the substituent, in contrast to the cyanates and thiocyanates due to the proximity of the nitrogen to the substituent and the chemical shifts observed in these compounds are also markedly different.

10.19   Nitrogen-Bearing Heterocycles

This section by its nature will contain many compounds but will, of course, be by no means comprehensive. Hopefully, it will help establish some degree of predictability in this very large category. Simple, small compounds will be listed first with a few larger heterocycles described later. All shifts are approximate.

N1, 156 ppm.

N1, 150 ppm.

If equivalent, N1/2, 248 ppm.

N1, 205 ppm. N2, 305 ppm.

N1/3 (equivalent), 242 ppm.

N1, 163 ppm. N3, 255 ppm.

N1, 250 ppm. N2/5, 310ppm.

N1, 248 ppm. N2/5, 330 ppm.

N1, 237 ppm. N2, 364 ppm. N3, 352 ppm.

N1/2, 370 ppm. N3/5, 280 ppm.

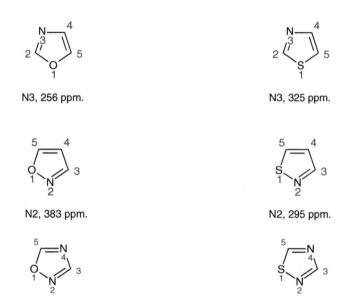

N1, 230. N2, 369. N3, 393. N4, 330 ppm.   N1, 278. N2, 379. N3, 333. N5, 307 ppm.

General observations: Substitution of an aryl ring for a methyl group adds approx. 10–20 ppm to shift of N1 nitrogen with minimal influence on others.

N3, 256 ppm.   N3, 325 ppm.

N2, 383 ppm.   N2, 295 ppm.

N2, 360 ppm. N4, 240 ppm.   N2, 275 ppm. N4, 310 ppm.

General observations: N-methylation on nitrogens (where possible) gives rise to a slight upfield shift though this trend is reversed in the case of alkyl chains.

Tautomerism may render seemingly distinct nitrogens equivalent. The true picture of chemical shifts only becomes clear on alkylation.

Nitrogens flanked by other nitrogens are likely to have significantly greater chemical shifts than those at the edge of the cluster.

In heterocycles where two nitrogens are joined, a nitrogen bearing a formal double-bond will always be at a considerably lower field than its singly-bonded partner.

Oxygen and sulfur have very different influences in five-membered heterocycles. An oxygen next to a nitrogen increases its chemical shift while a sulfur decreases it. This trend is reversed where the oxygen/sulfur is 'meta' to the nitrogen.

N1, 315 ppm.

N1/2, 400 ppm.

N1/3, 295 ppm.

N1/4, 333 ppm.

N1/3, 390ppm, N2, 440 ppm. (estimate only)

N1, 420 ppm. N2, 382 ppm. N4, 318 ppm.

General observations: (See also Section 10.23 for information on pyridine N-oxide.) The chemical shifts of heterocyclic nitrogens (where basic enough to accept a charge) are extremely sensitive to protonation. In the case of pyridine, full protonation gives rise to a chemical shift of approx. 210 ppm. Since protonation with a strong acid can be considered to be 100%, in solution, the ^{15}N shift can be used as a guide to estimate the amount of acid present in the case of partial salts. Formation of quaternary salts are similarly deshielding, N-Me pyridine giving a ^{15}N shift of approx. 200 ppm.

Alkyl (C-substituted) pyridines show little variation from the parent compound though lone-pair donating species (i.e. –OR and –NR2) are a special case and will be dealt with in Section 10.28 under 'Tautomerism in ^{15}N NMR'. Aryl (C-substituted) pyridines show a slight decrease in ^{15}N shift but this is only of the order of 5 ppm with no notable distinction between ortho-, meta- and para compounds.

N1, 313 ppm.

N2, 310 ppm.

N1/9, 315 ppm.

N1/9, 315 ppm. Note: Chemical shifts of compounds containing further nitrogens are analogous to those found in six-membered rings.

N1, approx. 133 ppm.

N1, approx. 133 ppm. Note: –N-methylation has little impact on ^{15}N shift but as in previous groups, alkylation with alkyl chains, branched chains and conjugated chains brings about a progressive downfield trend in chemical shift. As in the above case, chemical shifts of compounds containing further nitrogens can be roughly estimated from smaller, simpler heterocycles.

N1, 306 ppm.

N1, 304 ppm.

General observations: ^{15}N chemical shifts, in common with ^{13}C shifts, appear to be driven very much by electronic influence rather than anisotropy as the above examples demonstrate. The lack of any material difference in the shifts of the nitrogens in these two, related compounds exemplify this. (Note that in contrast, H12 in the proton spectrum of the second compound above would be expected to be much the lowest field of the proton signals, deshielded heavily 'through-space' by the proximity of the pyridine ring).

10.20 Nitriles

Alkyl nitriles are exemplified by acetonitrile itself, which has a well-established ^{15}N chemical shift of approx. 245 ppm. There is very little variation to be noted with longer alkyl chains but conjugation does depress the chemical shift considerably. For example, the shift of N7 in the compound below has been reported at approx. 270 ppm.

In the case of aryl nitriles (though rarely observed on account of, at the least, a 4-bond separation between the closest proton and the nitryl nitrogen) a small downfield shift is noted with a shift of approx. 257 ppm. Data indicates that strongly electron withdrawing groups in ortho or para positions on the aryl ring may contribute an additional 10 ppm to the nitrile ^{15}N chemical shift.

General observations: While ^{15}N NMR would never be considered to be the most favourable technique for the positive identification of nitrile compounds (^{13}C data is easier and quicker to acquire and the –CN carbon always falls in a well-defined area of the ^{13}C spectrum), it is none-theless reassuring to observe the nitrile nitrogen where possible.

10.21   Nitro Compounds

Alkyl nitro compounds are exemplified by nitromethane which has a well-established ^{15}N chemical shift of approx. 370 ppm. Longer alkyl chains depress the ^{15}N shift of the nitro group by a further (approx.) 20 ppm and branching at the point of attachment, an additional 5 ppm. The ^{15}N shift of the nitro compound shown below has been recorded at 377 ppm, indicating that a slight downfield shift can be expected in conjugated examples (R = alkyl chain).

In the case of aryl nitro compounds, the ^{15}N shift of the nitro group shows little discernible change with nitrobenzene giving a shift very close to that of acetonitrile (370 ppm). Substitution at ortho and para positions on the aryl ring also have no discernible influence on the shift of the nitro group.

   General observations: Correlations to the nitrogen of nitro groups can be a little weaker than average, so an increased number of scans may be needed to observe them, particularly in the case of weaker samples.

10.22   Nitroso and N-Nitroso Compounds

Nitroso compounds are associated with the most extreme ^{15}N shifts encountered. Alkyl nitroso compounds can be expected as low as 970 ppm (!) while the simple aryl nitroso compound below gives a shift of approx. 910 ppm.

In the case of the important N-nitroso compounds, chemical shifts can be expected in the region of 550 ppm. A compound of the type shown below is a very useful example as it demonstrates not only the chemical shift of the nitroso nitrogen but that of the nitrogen it is attached to. (R = a branched hydrocarbon, attached at the branch point.)

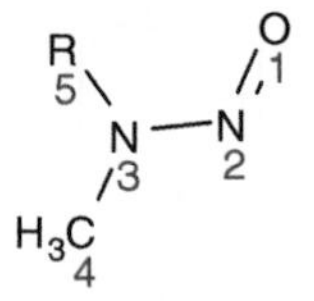

N3, 247 ppm. N2, 539 ppm.

General observations: A lack of a wide range of examples in these categories is not a significant problem as the extreme chemical shifts of both nitroso and N-nitroso species are fairly unique and most unlikely to be confused with any other nitrogen species. Care must be taken in dealing with such compounds, however, as signal folding (or possibly double folding!) will occur when typical sweep widths are used.

   Note that N-Nitroso compounds are known to be highly carcinogenic. Extra safety measures should be in place when handling them.

10.23   N-Oxides

In alkyl amines, N-oxide formation brings about a significant downfield shift when compared with the parent amine. The ^{15}N shift of trimethylamine N-oxide shown below is approx. 113 ppm.

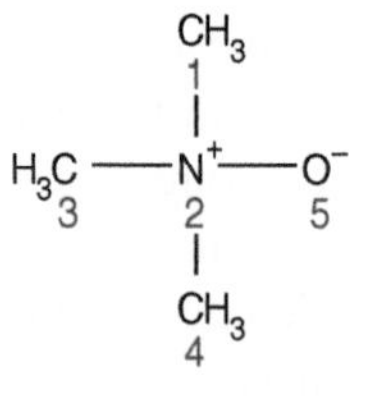

In the case of analogues featuring alkyl chains, there is very little additional deshielding. Exchanging one of the alkyl substituents for an aryl function also has little impact on the nitrogen shift though there is a lack of good examples to fully investigate this. A further example shown below demonstrates that the ^{15}N shifts of alkyl N-oxides are consistent. A shift of 115 ppm was recorded for N4 with only 2-bond correlations noted.

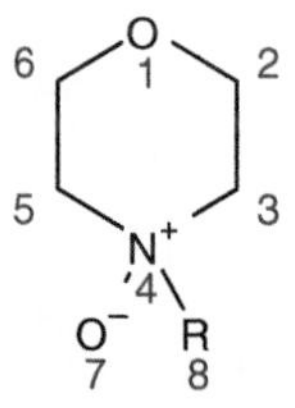

(R = alkyl chain)

The heterocyclic N-oxides are an important group of these compounds and are well exemplified by pyridine. (See also, Section 10.19.) Pyridine N-oxide has a nitrogen shift of approx. 294 ppm, compared to approx. 313 ppm for pyridine itself.

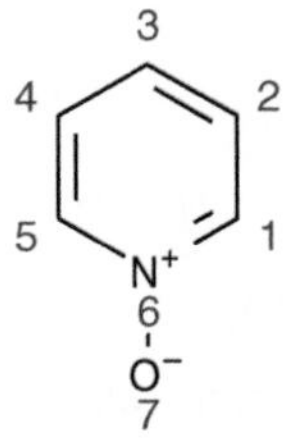

Another interesting compound is shown below.

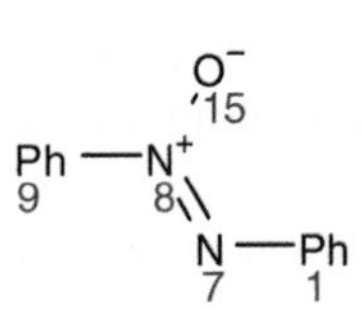

The N-oxide of a diazo compound showed correlations to the N-oxide, N8 (326 ppm) and as an internal comparator, N7 (340 ppm). This upfield shift is entirely comparable to those observed in pyridine N-oxides and other related heterocycles.

General observations: The chemical shift change seen in N-oxide formation is large in the case of alkyl compounds but far less so for heterocyclic compounds where the nitrogen is part of the ring system (e.g. pyridine). Nevertheless, careful comparisons of shifts with those of known parent compounds should enable the detection of N-oxides of this type.

It is worth noting that whereas in pyridine itself, 2-bond correlations to the nitrogen are invariably stronger than 3-bond correlations, in the pyridine N-oxides, the reverse is the case with 3-bond couplings proving to be much stronger than the 2-bond correlations. This observation can be a useful diagnostic tool in cases where the ^{15}N chemical shift is ambiguous. It is in notable contrast to the morpholine N-oxide cited earlier where only 2-bond correlations were reported. These phenomena reflect variations in the size of the ^{1}H–^{15}N couplings associated with the groups in question.

10.24   Oximes

Oxime nitrogens absorb at very low field, the di-methyl compound below, showing a shift of approx. 342 ppm.

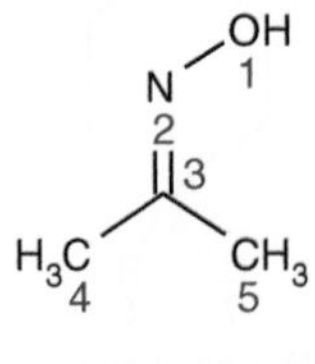

A cyclic analogue, shown below, showed a very similar chemical shift of 345 ppm for N7, with correlations from H3/5 and 8.

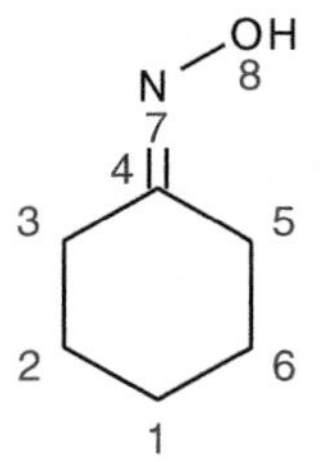

Oxime chemical shifts are further suppressed by conjugation as can be seen in the example below. In this example, H3 was shown to correlate to N9 (356 ppm) and H6 to N11 (358 ppm).

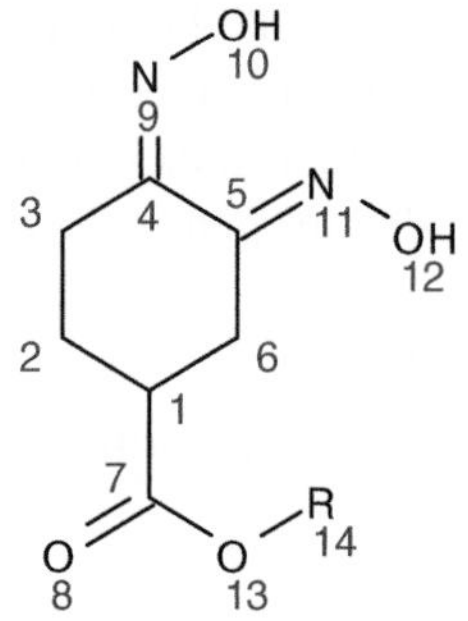

In the case of aryl oximes, the trans aryl oxime shown below gave a ^{15}N shift of approx. 365 ppm. Only one cis analogue was noted with a very similar shift.

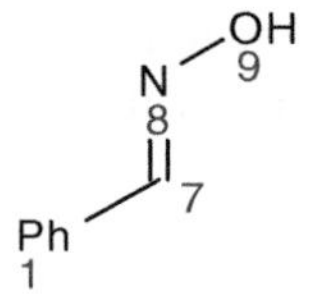

Di-arylation reverses the downward trend with the compound below showing a ^{15}N shift of approx. 346 ppm. This could reflect reduced conjugation of the aryl rings with the oxime bond caused by steric interaction between the two aryl rings.

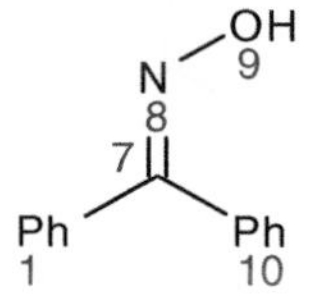

10.25   Sulfonamides

The sulfonamide nitrogen in the simple primary sulfonamide shown below has a ^{15}N shift of approx. 94 ppm. An aryl group in place of the methyl in this compound has virtually no effect on the sulfonamide nitrogen shift.

$$H_2N \overset{\overset{\displaystyle O}{\overset{2}{\|}}}{\underset{\underset{\displaystyle O}{\underset{3}{\|}}}{\underset{1}{S}}} CH_3$$
$$_4 \phantom{\overset{S}{1}}_5$$

Mono-methylation of the nitrogen moves it up to 81 ppm while the di-methylated compound gives a very similar shift (80 ppm). Alkylation and di-alkylation with alkyl chains give rise to the expected increase in chemical shift (approx. 99 ppm for both mono- and di-alkylated species). Mono-alkylation with a branched hydrocarbon (isopropyl) gives a further increase in chemical shift to approx. 110 ppm, while di-alkylation with this branched group produces no further change in chemical shift. It can be assumed that conjugated species will be further deshielded as with other groups though there is a lack of useful examples to underpin this.

A single aryl group on the sulfonamide nitrogen (below) gives a shift of approx. 122 ppm.

$$\underset{1}{Ph} \diagdown \underset{7}{NH} \overset{\overset{\displaystyle O}{\overset{9}{\|}}}{\underset{\underset{\displaystyle O}{\underset{10}{\|}}}{\underset{8}{S}}} \underset{11}{CH_3}$$

A second aryl group reverses the trend, giving a shift of 99 ppm.

General observations: Trends observed in the sulfonamides tend to mirror those seen in the amides. Substitution on the sulfur side of the moiety has little impact on the chemical shift of the nitrogen, while direct substitution on the nitrogen can have a significant impact. Note that correlation from protons across the sulfur is uncommon when using standard acquisition parameters.

10.26   Ureas and Thioureas

Urea itself (shown below) has a ^{15}N chemical shift of approx. 77 ppm.

$$\overset{\overset{\displaystyle O}{\overset{1}{\|}}_2}{\underset{\underset{H_2N}{4}\qquad \underset{3}{NH_2}}{\diagup\ \diagdown}}$$

Mono and di-methylation bring about a small but definite upfield shift of approx. 5 ppm, while the addition of a single alkyl chain depresses chemical shift of the alkylated nitrogen to approx. 88 ppm. A second alkyl chain deshields the nitrogen further to approx. 96 ppm. In the case of branched hydrocarbons, a further deshielding may be expected, the di-isopropyl urea (one isopropyl group on each nitrogen) giving a shift of 101 ppm. Conjugation can be expected to depress chemical shift considerably but there are very few representative examples available.

A single aryl group brings the arylated nitrogen down to around 106 ppm. No bi-aryl compounds available for comparison.

Thioureas make for an interesting comparison. Thiourea itself has a shift of approx. 108 ppm. Mono-methylation gives an upfield shift to approx. 101 ppm while di-methylation lifts it further to approx. 95 ppm. Mono-alkylation with an alkyl chain depresses chemical shift to approx. 120 ppm while di-alkylation appears to have little extra discernible effect.

While mono-arylation gives a shift of 130 ppm, there were no useful examples of bi-arylated (on the same nitrogen) compounds available for comparison. The symmetrical bi-arylated compound shown below also showed a chemical shift of 130 ppm.

General observations: Ureas and thioureas behave in a broadly similar fashion in terms of their chemical shifts but the thiourea species will be likely to give ^{15}N shifts approx. 30 ppm lower than the corresponding urea.

10.27   Other Unusual Compounds

(Note: No attempt has been made to categorise compounds shown in this section. They are presented purely in the chronological order they were encountered.)

In the compound above, featuring an N-B moiety, H1 gave a 1-bond correlation to N1 (135 ppm) and a 3-bond correlation to N5 (106 ppm). Conversely, H5 gave a 1-bond correlation to N5 and a 3-bond correlation to N1.

Another boron–nitrogen group is shown in the compound below. N4 showed correlations from H3 and H5 (48 ppm).

In the compound below, H3/4 gave correlations to N1/2 (159 ppm).

N11 in the conjugated hydroxamic acid ester below, N11 showed correlations from H13 and 14 (188 ppm).

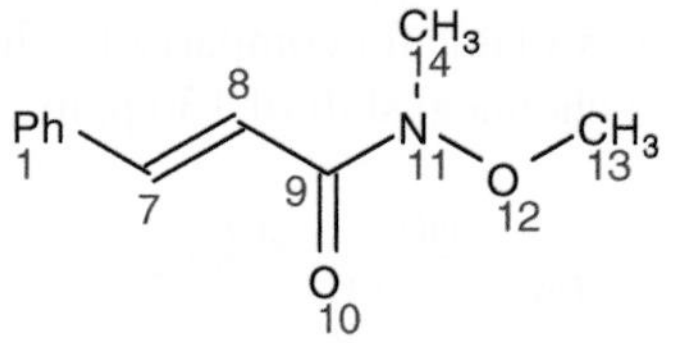

In the compound of the type shown below, N8 showed correlations from H8 and H7 (90 ppm) while H12 and H14 showed correlations to N12 (84 ppm).

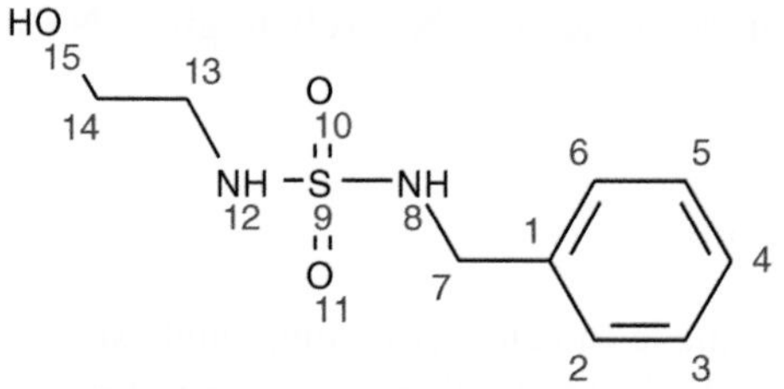

In the imine-like compound below, N7 showed correlations to H3/5 (302 ppm).

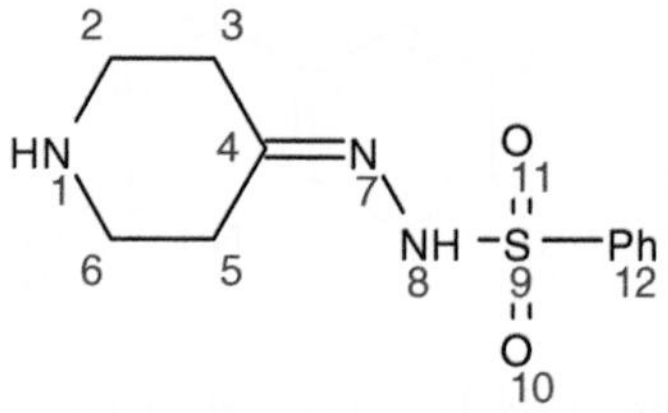

In the compound below, N7 showed correlations to the ortho aryl protons and H8 (104 ppm) while N8 showed a 1-bond correlation only from H8 (153 ppm).

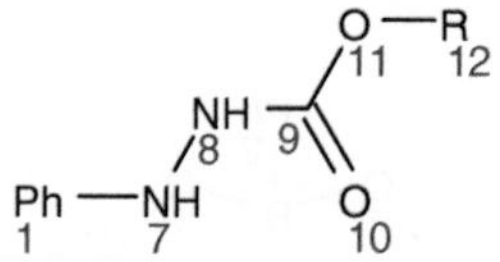

In the unusual compound below, N10 showed a correlation to H13/14 (156 ppm).

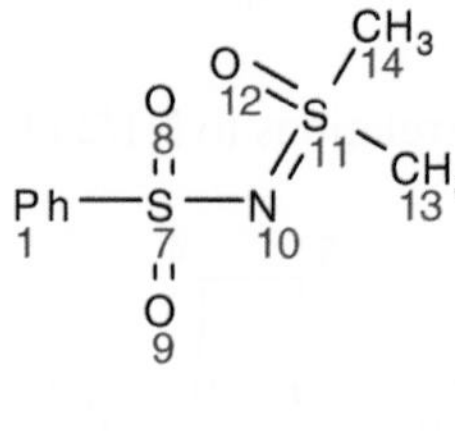

N10 in the compound below showed correlation from H15 (424 ppm).

In a compound of the type shown below, N2 showed correlations to H9 and 10 (179 ppm).

In the compound below, N9 showed correlations from H7, 8 and 9 (96 ppm).

In the isomeric compound shown below, N9 was noted at a somewhat deshielded position (107 ppm), showing correlations to H7 and 8.

In a compound of the type shown below, the charge on the quaternary imine was seen to exert a very large upfield influence on the nitrogen shift of N7.

H2/5 showed a correlation to N7 (97 ppm!). This follows the trend seen in heterocycles where a nitrogen is protonated or quaternized. (N9/11 showed expected correlations from methyl groups, H12 etc., 79 ppm.)

In a compound of the type shown below, H26 showed a correlation to N6 (176 ppm) and is clearly deshielded by N7 which itself correlates to H9/13 (312 ppm). (N11 also correlates with H9/13 and shows a shift of 87 ppm.)

10.28 ^{15}N Topics

10.28.1   1-, 2-, 3- and 4-bond Correlations

1-bond correlations in ^{15}N HMBC spectra are relatively commonplace and can often provide useful information for confirming the identity of exchangeables. Any reasonably sharp NH proton signal can show a 1-bond correlation (e.g. amides, carbamates, ureas, etc.) though this observation tends to be a little less common in the case of amines as amine NHs are often inherently broader and so less able to correlate.

2-bond correlations are often observed though for most classes of compound, they are usually weaker than 3-bond correlations (in the context of an HMBC experiment optimized for 7 Hz couplings) with a few notable exceptions. For example, as has been noted previously in Section 10.23 (see *'General observations'*), in the case of pyridine and related heterocycles, 2-bond correlations are stronger than 3-bond correlations. In some compounds, 2-bond correlations are not observed at all even where the sample is relatively concentrated and the signal-to-noise is excellent. This can lead to confusion as in an attempt to establish a weak 2-bond correlation, weak (but relatively strong!) 4-bond correlations may be observed. This can be more of an issue in cases where the experiment has been optimised for smaller ^{1}H–^{15}N couplings.

3-bond correlations are in most cases likely to be the strongest observed. However, there are cases where 3-bond correlations can be weaker than expected. 'Cis' correlations often fall into this category as, for example, the weak correlation observed between H7 and N1 in a compound of the type shown below (234 ppm), though relatively strong 2-bond correlations were in evidence for H2–N3 (168 ppm) and for H2–N1.

Note also the lack of correlation seen from H4 to N2 and from H7/11 to N1 in a compound of the type shown below, despite reasonably strong correlations between H13 and N1 (213 ppm) and between H12 to N2 (303 ppm). Phenomena of this type are quite commonplace and can easily lead to errors in interpretation. Looked at in a simplistic manner, we have to accept that while some nitrogens can be detected quite easily, others which appear to enjoy a similar spatial arrangement with the correlating protons, may not show any correlation at all in the same spectrum. This might appear to be totally at odds with what we have come to expect from our early experiences with 1-D proton spectra, where we learnt that all protons in a molecule can be expected to present themselves in the spectrum. But indirect detection in a 2-D experiment doesn't work like that. We are not observing the nitrogens directly. We are detecting the coupling between a proton or protons and a nitrogen nucleus. The response, or sensitivity of the experiment, is dependent on the size of the proton–nitrogen coupling that the experiment has been optimised for. Just because you don't see a correlation that you might reasonably expect, you must not fall into the trap of concluding that the no-show nitrogen in question isn't there at all! These observations are of course just as applicable to ^{13}C HMBC spectra but they are considerably more problematic in the case of the ^{15}N experiment as our signal-to-noise may well be more marginal, given the extremely low natural abundance of the ^{15}N nucleus and also because we don't have the luxury of a 1-D spectrum to check it against.

4-bond couplings, though generally unusual, can sometimes be seen where a 'trans' arrangement exists between proton and nitrogen. For example, in a compound of the type shown below, a 4-bond correlation was noted between H8 and N1 (253 ppm).

Such couplings are usually weak and care must be taken to avoid confusion with 2-bond correlations as noted earlier.

General observations: In all cases, the observation of a correlation, or the lack of it, is dependent on the size of the ^{1}H–^{15}N coupling in relation to the size of this coupling that has been optimised for in the experiment. As the actual values of these couplings are unlikely to be known, particularly in cases where nitrogen atoms are in unusual environments, the best initial approach can only be to ensure as good a signal-to-noise ratio as possible within the context of available instrument time.

10.28.2   'Through-Space' Correlations

A very interesting aspect of ^{15}N HMBC spectroscopy is the potential for observing long-range 'through-space' correlations. These may be encountered in compounds where the nitrogen and correlating proton(s) are held in close proximity to each other by steric constraints. In a compound of the type shown below, the rigid heterocycle and steric constraints imposed by the substitution of the amine and the alkene moieties, force the lone pair of N16 into alignment with the H13 protons (CH$_2$) which correlate 'through-space' (91 ppm).

In other examples, like the one shown below, it has been noted that sufficient rigidity may be found in molecules where conjugation encourages a planar conformation, to facilitate through-space correlations.

In this case, H7 correlated strongly to N2 (296 ppm), though it did not show a correlation to N1. (Note that the assignment was confirmed by the correlation seen from H15 to N1 [233 ppm].)

In some cases, H-bonding can initiate 'through-space' correlation as in the example below.

In this case, H25 is positioned to H-bond to N17 and correlates to it (230 ppm).

10.28.3 Tautomerism in ^{15}N NMR

Tautomerism can have two important consequences for ^{15}N observations. As with other nuclei, the observations made will always depend on the 'NMR timescale' of the tautomeric process taking place but in cases where two tautomers are similarly energetically favoured (such as those shown below), it may not be possible to observe any correlations in the nitrogen domain.

This is because the nitrogen absorbances can be greatly broadened by the tautomeric process and this lack of sharp nitrogen resonances will inhibit the correlations from the protons which we rely on to observe the nitrogens. (The tautomerism may well broaden proton resonances as well, which will further inhibit correlations to nitrogens.)

Another interesting consequence of tautomerism is the possibility of being able to determine the structure of a preferred tautomer in a given solvent. While this may not be too surprising in the case of pyridines substituted in the 2′ position with –OH or –NH2 groups, some examples may be less obvious such as the one shown below. H27 showed a strong correlation to N24 which had a chemical shift of 171 ppm. This was at least 100 ppm higher than expected and indicated that the preferred tautomer was in fact the structure shown next to it. This structure was supported by the size of the proton–proton coupling between H25 and H26 (7.1 Hz) and also, by the ^{13}C shift of C21 (162.6 ppm).

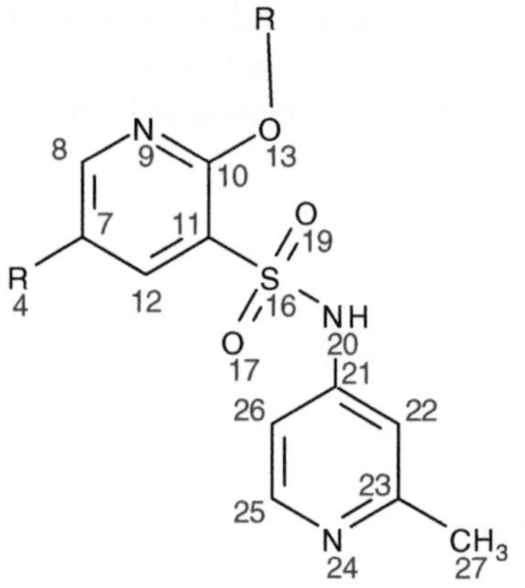

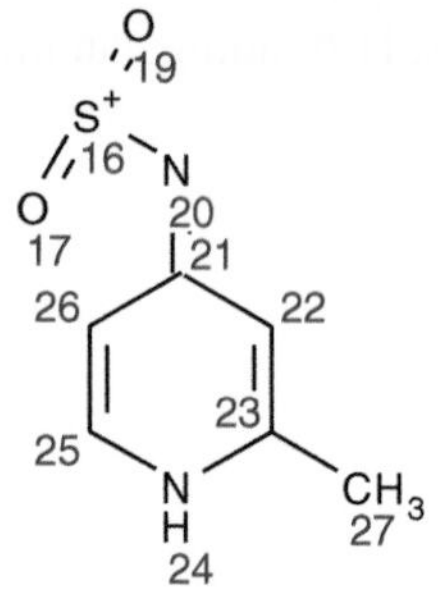

Preferred tautomer

10.28.4   Restricted Rotation

As with other nuclei, restricted rotation can have consequence in the context of ^{15}N NMR. NMR timescale aside, it is sometimes possible to observe distinct signals for nitrogen atoms that are close to a site of restricted rotation, though the chemical shift difference between them is often small and can be on the limits of resolution in the ^{15}N domain when using typical acquisition parameters. An example of such a small rotameric shift difference is shown below. (Restricted rotation is brought about by donation of the lone pair on N8 into the heterocyclic ring, i.e. C4–N8 bond has some double-bond character.)

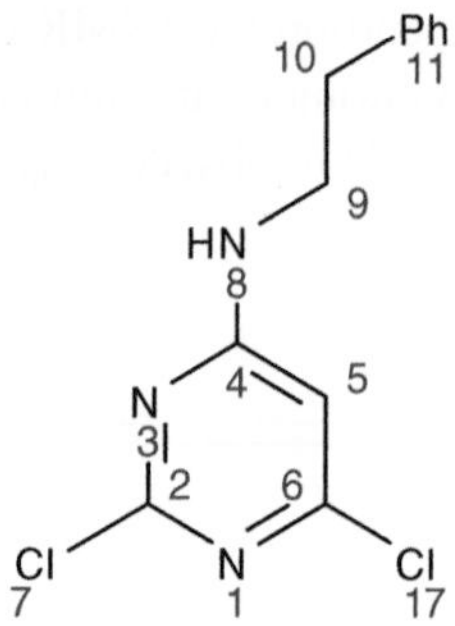

In this compound, H9 and 10 were both seen to correlate to N8 which showed two distinct but very close shifts of 98.8 ppm (major rotamer) and 100.3 ppm (minor rotamer). N3 showed correlations from H5 and H8 but no rotameric differentiation. Accidental equivalence of the two rotameric species for this nitrogen can be assumed.

10.28.5   Protonation and Zwitterions

As already discussed in earlier sections, protonation does have an impact on ^{15}N chemical shifts; marginal in the case of alkyl amines but large in the case of N-bearing heterocycles (where nitrogen is part of the heterocyclic structure and basic enough to accept a charge). These observations can be useful in studying protonation in different solvents and in determining the charge distribution in zwitterionic compounds.

The two related compounds shown below are particularly interesting in this regard.

In the case of the ethyl ester, H1 and H3 showed an unremarkable correlation to N1 (74 ppm) while H11 showed an equally typical correlation to N7 (262 ppm). The analogous carboxylic acid, however, showed a corresponding H3–N1 correlation at 89 ppm and an H11–N7 correlation at 168 ppm! The explanation for these discrepancies between the two similar compounds can be explained in terms of internal protonation in the case of the carboxylic acid. While the ability to form a six-membered H-bonded ring (H13–N7) may well enhance internal protonation, a study of a similar pair of compounds, selected to prevent the formation of such a H-bond, yielded very similar shifts in both ester and carboxylic acid species. (Note that all spectra in this study were acquired in CDCl₃ solution and that spectra acquired in DMSO did not show such protonation in solution.) As the chemical shift of N7 would hardly be any different in the presence of an acid, it can be reasoned that the protonation occurs exclusively at N7 and that the associated downfield shift of N1 reflects the re-distribution of charge within the molecule in response to this.

As the range of chemical shifts found between protonated and un-protonated heterocyclic nitrogens is large (~100 ppm), the observation of ^{15}N shifts in such molecules can be used as quite a precise tool for establishing protonation in solution. For example, the ether/alcohol analogues (below) of the ester/acid pair above also showed a corresponding shift. While the shift in this case was much smaller, it was still large enough to be significant.

In the methyl ether, shifts of 261 ppm and 73 ppm were recorded for N7 and N1 while in the alcohol, shifts of 254 and 74 ppm were noted. Though small, this difference reflects a small degree of protonation of N7 by the alcohol –OH (13).

We hope that this chapter has given a reasonable insight into the whole ^{15}N NMR subject area and that we have given a fair indication of the potential of the technique as a front-line problem-solving weapon.

11

Some Other Techniques and Nuclei

If this chapter appears to be a bit of a 'dog's breakfast', we can only apologise! It covers a number of topics that didn't sit happily in other sections of the book and that we didn't wish to extend into further chapters, so please bear with us…

We have already covered what in our view are the most important and relevant techniques – but there are many others you may have heard of, all with their own enticing sounding acronyms. We'll now take a look at some of them and try to outline some of their potential uses as well as their shortcomings.

11.1 HPLC-NMR

Linking techniques together might seem like a good idea in theory but, in practice, there can be as many problems as potential advantages. HPLC-NMR does have undeniable use in the field of bio-fluid NMR and in process control in a production environment but we feel that it has little to offer the organic chemist looking to monitor a reaction.

The two techniques don't really sit happily together. High Performance Liquid Chromatography (HPLC) is essentially a dynamic technique – solvent and solute move continuously from injection port through column to detector. Data acquisition takes place rapidly as the various fractions pass through the detector. But NMR detection isn't like that. It takes time – and with dilute solutions, the sample just isn't resident in the flow cell (a flow cell replaces the conventional NMR tube in HPLC-NMR) for long enough for any useful acquisition to take place. In order to overcome this problem, it is possible to use a 'stop-flow' technique where the flow is stopped and the fraction contributing to a peak is 'parked' in the probe for long enough to acquire some useful data. Of course, this can lead to serious chromatographic problems. It is quite possible that while the early-running peaks are in the flow cell, the later running ones are still on the column. Stopping the flow at this stage will inevitably lead to fractions broadening themselves by diffusion on the column.

Another serious problem is that of the chromatography solvent gradient. It is common practice for reverse-phase columns to be run using a solvent gradient system so that polarity of the solvent is gradually changed from polar to non-polar throughout the separation by altering the

---

Essential Practical NMR for Organic Chemistry, Second Edition. S.A. Richards and J.C. Hollerton.
© 2023 John Wiley & Sons Ltd. Published 2023 by John Wiley & Sons Ltd.

ratio of two different solvents during the run. This has the benefit of extending the range of polarities that can be accommodated on one column within a sensible run time. The problem with this, from an NMR perspective, is that shimming, and indeed probe tuning, will be very much altered as the run progresses giving rise to line-shape and sensitivity issues. Note that even without this complication, resolution in a flow cell can never be as good as in a conventional tube as a flow cell can't spin! The solvents themselves are yet another issue. To run such a system on deuterated solvents would be prohibitively expensive and so normal non-deuterated ones are used. This of course gives rise to massive solvent peaks which must be suppressed, denying you access to potentially important parts of the spectrum.

There are numerous other problems associated with the technique. Such systems need very careful setting up to ensure that the fractions park accurately in the flow cell so as to maximise concentration and hence signal-to-noise. Other minor irritants can include various plumbing problems, blockages causing capillaries to burst off, wet carpets, etc.

There are several variations on the theme of instrument set-up which have been used in an attempt to overcome the shortcomings inherent in the concept. For example, as an alternative to the stop-flow method, the various fractions can be collected into sample 'loops' (small loops of capillary tubing) which can then be flushed into the flow cell and studied at leisure. After spectroscopic examination, each sample can then be returned to its loop and the next pumped in. Fractions suffer dilution in this way but this approach would seem to offer an advantage over stop-flow in that at least the chromatography is not compromised by diffusion on the column.

Another variation is that of trapping eluting compounds onto solid-phase cartridges and then washing them off as required using a suitable *deuterated* solvent. In this way, the problem of solvent suppression needed for dealing with non-deuterated solvents is neatly side-stepped. But all of a sudden, it would seem that the two techniques are becoming more and more segregated again. And there's the rub – HPLC is an excellent technique and so is NMR but what is good for one is bad for the other and vice versa. Perhaps then, it is best if we do not force them into an uneasy alliance. By analogy, various attempts have been made in the past to build an amphibious car but results have generally been characterised by mediocre performance on land and worse on water. Engineering to cater for many disparate requirements in any system can lead to compromises which adversely affect the performance of the whole system.

11.2   Flow NMR

In a sense, flow NMR is like HPLC-NMR without the chromatography part. It has found use in the field of 'Array Chemistry' where 'libraries' of compounds, usually with a common motif, are made – or at least presented – in the wells of a 'plate'.

Systems have been developed by some of the major spectrometer manufacturers to deal specifically with this type of application. These systems are designed with automation very much a priority. Typically, an integrated robot adds a predetermined volume of solvent to each of the wells and then injects the resultant solution into a flow line that transfers it into the spectrometer's probe, which is of course fitted with a flow cell. Spectroscopy can then be performed without the time constraints of the HPLC-NMR system and the sample returned to the well on the plate where it came from, or into a fresh one if required.

With careful fettling, these systems can work quite well but they are not without their potential pitfalls. For example, there is always a danger that the samples will suffer some degree of cross-contamination as they are all being drawn up by the same automated syringe and transported through the same capillaries. Obviously, such systems use a flushing cycle between samples but it is not impossible for a particularly 'sticky' sample to hang around in some recess of the plumbing only to be gradually flushed out with the passing of subsequent samples. So it is not reasonable to expect the re-formatted plate to be of as good a quality as it was before spectroscopic investigation. There is also the question of sample recovery and dilution. It would be unreasonable to expect 100% sample recovery after shunting a solution through several metres of plumbing and recovery rates will vary with individual system set-up. These factors can have implications if you wish to re-visit one of the samples of the library for further investigation. Ideally, you would do any further investigation when the sample in question is in the probe but there's not much point in having an automated system if you have to stand over it all the time!

Obviously, flow NMR can generate a huge amount of data. Library plates can often hold 96 samples and an over-night run can easily present you with 96 spectra to look at the following morning. In other words, such a system could generate spectra faster than you could interpret them. In these circumstances, a rather cut-down approach to interpretation is required. The chemistry under investigation might, for example, be that of the reaction of a specific amine with 96 different carboxylic acids. Rather than address every minute feature of every spectrum, you might have to make do with an indication that the reactants have reacted as desired and an overall impression of sample purity.

Flow NMR has recently been eclipsed by the advent of robotic sample handling systems capable of dealing with very small sample quantities and volumes. We now have a system operating in our lab that makes up samples directly into 1 mm NMR tubes, using only about 8 microlitres of solvent. These can be run under automation and the tubes emptied back into the plate wells by the same robot. This technology offers superior performance and largely gets around the problems of contamination and recovery.

11.3   Solvent Suppression

When dealing with high-quality samples, solvent suppression is not an issue that should ever cause concern. However, if for some reason, your sample is heavily laden with some solvent or water which cannot be easily removed from your sample, then you might need to consider some form of solvent suppression. Why? Because when acquiring the spectrum of a sample that contains a peak or peaks that are orders of magnitude larger than those of interest, the receiver gain requirements will be set to cater for the large peak(s) at the expense of the small ones. This would be analogous to acquiring with far too low a receiver gain and yield very poor signal-to-noise for the peaks of interest.

Such a problem can be addressed by artificially suppressing the huge unwanted peak(s) so that the smaller desired peaks can be acquired optimally. At its crudest, peak suppression is nothing more than a decoupler signal of suitable power, centred on the unwanted peak. Once the problem peak is saturated to oblivion, the rest of the spectrum can acquire normally. This method works well enough in most cases but far more subtle methods have been developed. It is possible, by using a simple macro program, to move the decoupler signal rapidly back and forth over the peak to be suppressed. This can improve the resultant spectrum by minimising suppression artefacts.

Other even more cunning methods have been devised to suppress the water signal in samples that have a large water content (e.g. bio-fluid samples) such as the WET and the WATERGATE pulse sequences. Other sequences have been devised to cope with signals from carbon-bound hydrogens. Some of these actually collapse the ^{13}C satellites into the main ^{12}C peak prior to suppression. Such a sequence would be useful if you were forced to acquire a spectrum in a non-deuterated solvent.

11.4   MAS (Magic Angle Spinning) NMR

Synthesis of compounds on solid-phase supports became quite popular in the late nineties and though interest might have waned a little more recently, there may still be a demand for it in some establishments. If monitoring reactions carried out on resins is what is required, then a MAS probe is the only way to go.

The NMR of solids is a specialist field and as it is of little or no relevance to the organic chemist, is outside the scope of this book and so we will say very little about it. The main problem associated with solids is that the lattice relaxation is very efficient causing NMR lines to be extremely wide. (Remember: The faster the relaxation time of a nucleus, the broader its NMR peak will be.) Spinning samples at very high frequency (2–6 kHz) at the so-called magic angle helps to minimise this broadening because it sharpens NMR lines by negating the effects of chemical shift anisotropy which arises (in solids and semi-solids) as a result of the directional character of chemical shifts with respect to the applied magnetic field. Chemical shift anisotropy (and dipolar interactions, another source of broadening) vary with the term, $3\cos^2 q-1$ where 'q' is the angle from the vertical axis of the applied field. This term becomes zero when q is 54° 44' – the 'magic' angle.

The problem with trying to run spectra of solid-phase gels with organic compounds bound to them is that the materials are, in NMR spectroscopy terms at least, 'solid-like'. Trying to run them in conventional probes is a complete waste of time. NMR line widths will be hundreds of Hz wide and no useful information will be forthcoming. Running them in a MAS probe can greatly improve matters – if the resin is of the right type. The key to achieving sharp lines is molecular mobility or, if you like, the removal of the very efficient lattice-relaxation pathway mentioned above.

Numerous resin supports are commercially available for solid-phase synthesis and some allow the acquisition of quite reasonable quality spectra of compounds bonded to them – and some don't. The resins to avoid (if you intend trying to monitor your reactions by MAS-NMR) are any that are based purely on cross-linked polystyrene. These are too rigid and afford little or no mobility to any bound compound. These resins are relatively cheap and have high specific loadings but will give very poor spectra even in a MAS probe. We see little point in running spectra of compounds on these resins as the quality of the spectra make them virtually useless – and perhaps worse – potentially misleading.

Compounds bound to resins such as Tentagel and Argogel, on the other hand, give spectra that can yield useful results. These resins are still polystyrene-based but differ in that they have long PEG (PolyEthylene Glycol) chains bonded to them and the compounds synthesised are bound to the end of these chains via a linker. These chains allow considerable freedom of movement at the end of the chain and thus the bound compound experiences something far more akin to a normal liquid environment. There are certain unique problems associated with the acquisition of spectra of compounds bound to solid-phase supports. One problem associated with these samples arises from the very long PEG (PolyEthylene Glycol) chains which connect the 'linkers' bearing the synthesised molecule, to the beads themselves. The PEG chains are about

50 units long giving a most unwelcome 200 proton signal at about 3.5–4 ppm. This often completely obscures important signals in this region and nothing can be done about it.

The broad polystyrene signals of the support beads are another major problem and render integration of most regions of the spectra impossible. These broad signals can be suppressed by using the CPMG (Carr–Purcell–Meiboom–Gill) pulse sequence. (This sequence works by multiple de-focusing and re-focusing of the sample magnetisation. The sharp lines re-focus and the broad ones do not.) Unfortunately, this does not mean that CPMG spectra can be integrated, as signal suppression occurs for all broad signals – irrespective of their origin and of the nature of the broadening. We recommend acquiring both the ordinary spectrum and the CPMG spectrum and looking at them alongside each other.

In practical terms, very little material is needed, as the MAS probe is extremely sensitive. More material tends to be lost in handling than actually ends up in the tiny NMR tubes used in this probe (typical volume, 40 microlitres). One mg of bead is usually ample. Note that this corresponds to far less than 1 mg of compound, of course. We have found that it is best to introduce resin-bound sample into the 'nano-tubes' before adding the solvent as wet resin is extremely difficult to handle. $CDCl_3$ or CD_2Cl_2 is used as the 'solvent', though its real purpose is to disperse the beads and cause them to swell, as well as to provide a source of deuterium for locking the spectrometer. This swelling function is important as it allows the PEG chains to move about more freely. The less restricted the motion of the chains, the sharper will be the signals from them.

Some MAS probes are single-coil, allowing proton-only acquisition, and some are dual-coil, allowing the acquisition of 2-D proton-carbon data. Note that MAS probes can be used for ordinary solution work and though very labour-intensive to use, can give excellent sensitivity where the available compound is limited and signal-to-noise is at a premium.

11.5 Pure Shift NMR

The idea of this experiment is to remove all proton–proton couplings to produce simple chemical shift information (Spectrum 11.1).

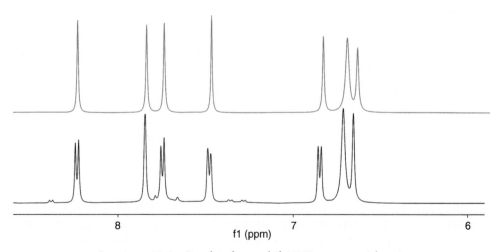

Spectrum 11.1 Simulated pure shift NMR spectrum (above).

There are many potential advantages to pure shift NMR in both 1-D and 2-D experiments. Losing the couplings can, in theory, increase the signal-to-noise although this can be lost due to signal losses during the experiment. It can also make the data easier for computers to interpret, having simple chemical shift information, not complicated by all those couplings. The pure shift pulse sequences can be quite demanding of amplifiers and probes depending on the approach used. We seldom use this technique because we find the coupling information useful but there is no doubt of its utility in strongly overlapped systems.

11.6 Other 2-D Techniques

11.6.1 INADEQUATE

High on any NMR spectroscopist's wish-list would be a technique that could be used to establish connectivities directly between carbon atoms. Such a technique does exist and it goes by the name of INADEQUATE (Incredible Natural Abundance Double Quantum Transfer Experiment). While this might sound fantastic in theory, everything in the garden is far from rosy…

In order for this to work, it is necessary to have molecules where adjacent carbons are both ^{13}C. Given that only 1.1% of the entire carbon content of any molecule (assuming no selective enrichment) is ^{13}C, then statistically, you will find adjacent ^{13}C atoms in only one molecule in about 10000! And this is the real problem with the technique – inadequate sensitivity. Here, we are talking about a method which has sensitivity so low that we would be needing at least 100 mg of material and still need many hours of scanning to get anything like a useable S/N.

Research Chemists, in our experience, seldom have this amount of material to play with but even if you are fortunate in this respect, solubility could well be an issue. Dissolving 100 mg of compound in 0.6 ml of solvent is seldom possible.

Practical constraints prevent this technique from living up to its potential, even in this, the era of the superconducting probe. Until sensitivity improves by at least another order of magnitude, the INADEQUATE experiment will remain just that – inadequate by name and by nature.

11.6.2 J-Resolved

Another useful-sounding technique is the Proton J-Resolved experiment in which chemical shift and coupling information are separated into two different dimensions. This is equivalent to turning the peaks sideways and looking down on them from above so that viewing them in the *x*-direction, they all appear as singlets but in the *y*-direction, they reveal their multiplicities.

This would be very useful indeed, particularly where overlapping multiplets are concerned. Unfortunately, in the very circumstances where the technique would be most useful, it tends to fall over with strong artefacts becoming intrusive in strongly coupled systems.

11.6.3 DOSY

Most of the approaches we have seen, rely on manipulations of nuclear spins. DOSY (Diffusion Ordered SpectroscopY) is a little different in that it is based on properties of the whole molecule. In this case, what we are measuring is the diffusion rate of a molecule. Normally this is used for mixtures so they can be resolved in the NMR tube. The technique works by using field gradients to make the sample experience a different field in different parts of the tube. If a molecule moves during the acquisition process, it will experience a different field. The more it moves, the more different the field it experiences. This has the effect of decreasing the intensity of the signals (the more they move, the more they are attenuated). If we change the strength of the gradient for each 2-D increment, acquire the data and FT the result, we end up with a typical 2-D plot (see Spectrum 11.2) where the chemical shift of the signals is shown on the *x*-axis and the diffusion rate on the *y*-axis. Obviously, because the diffusion rate is a whole molecule property, you see all the signals for the same molecule on the same horizontal line.

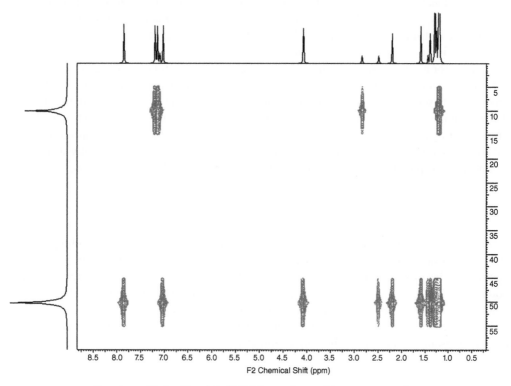

Spectrum 11.2   Simulated DOSY spectrum of two compounds.

   Fine? No need for chromatography then? Well, unfortunately it is not quite as easy as that. While the experiment has improved over the years, it still struggles to resolve compounds of a similar size and mobility. This means that your mixture of regioisomers will probably not resolve using DOSY. That said, some recent work on using micelles and shift reagents looks promising to improve the technique further and it may figure more prominently in the future.

11.7   3-D Techniques

If 2-D NMR techniques are really useful then 3-D ones must be even more so…shouldn't they? A number of 3-D experiments have been devised which are in fact produced by merging two, 2-D experiments together. The results could never be plotted in true 3-D format since etching them into an A3 sized block of glass would not be practical and viewing them as some sort of holographic projection would probably not be feasible! In essence, 3-D spectra have to be viewed as 'slices through the block' which effectively yield a series of 2-D experiments. It is possible to combine techniques to yield experiments such as the HMQC-COSY and the HSQC-TOCSY.

Of course, what works well on a 10% solution of ethyl benzene in 5 h may not be so good when you're confronted with an impure mg of dubious origin! These techniques may well be useful in specialised circumstances but are probably outside the realm of what a practicing organic chemist would want to get involved with.

We have tried to point you in the direction of the experiments that we have come to use and rely on, with good reason. There are dozens more out there that have been developed; some have evolved and are now generally known by another name (e.g. the ROESY experiment used to be known as CAMELSPIN) and some have been superseded and fallen by the wayside. If you have the instrument time and the inclination, by all means play but if time is of the essence, as it usually is, stick with the safe options.

And now for a brief look at a couple of the other potentially significant nuclei…

If you have suitable hardware, any of the 60-plus NMR-sensitive nuclei can be observed, though some are more suitable than others and in terms of organic synthesis, many are largely irrelevant. The most suitable nuclei for observation have three characteristics in common – high natural abundance (ensuring good sensitivity), a spin quantum number of ½ (ensuring that an uncoupled signal will appear as a singlet) and no quadrupole (ensuring that line shape will be naturally sharp). Both fluorine and phosphorus tick all three boxes, making them two of the more important 'also-rans'.

11.8 Fluorine (^{19}F) NMR

As we have already pointed out in the section dealing with heteronuclear coupling, it is not always necessary to confirm the presence of a particular heteroatom by acquiring the NMR spectrum of that nucleus. More often than not, the heteroatom will have a clear signature in the proton or carbon spectrum. Fluorine and phosphorus are both examples of nuclei that couple to protons and carbons over two, three, four and even more bonds.

There are cases where coupling from the heteroatom to neighbouring protons is not observed for some reason. Consider the following examples…

In terms of proton NMR, these two compounds would give remarkably similar spectra. The solitary fluorine on the 1,4 di-substituted ring would certainly couple to the protons both ortho- and meta-to it but the fluorines of the –CF$_3$ group would not show any discernible coupling to the ring protons and neither would the fluorines of the fully substituted ring. Yes, you could certainly discriminate between the two by ^{13}C NMR but if you only had a mg of each, you would be really struggling for sufficient signal-to-noise to observe the key carbons – even if you had a top-of-the-range spectrometer at your disposal. By ^{19}F NMR, however, the distinction would be easily made.

There are a number of features regarding ^{19}F NMR that are worthy of special note. Firstly, spectra may be acquired that are either proton-decoupled, or un-decoupled. We recommend acquiring both if you can. A comparison between the two can yield valuable information regarding neighbouring protons and other fluorines. Secondly, the range over which fluorine resonances occur is very large indeed and a sweep width of about 400 ppm (-300 to $+100$ relative to CFCl$_3$) is required to capture all the organo-fluorine resonances that can be expected. When such a large sweep width is plotted on a single sheet, it effectively compresses any couplings, making them almost unnoticeable. For this reason, expansion of peaks is recommended in coupled spectra.

Another consequence of the large sweep width needed for ^{19}F acquisition is that the electronics of the instrument are pushed to the limit: it is difficult to generate uniform RF irradiation over such a large frequency range and for this reason it may be necessary to acquire spectra in different spectral ranges, depending on the expected fluorine environment. This is particularly so in the case of high-field ($>$400 MHz) spectrometers.

There seems to be no universal reference standard in ^{19}F NMR as there is in proton NMR and this can cause confusion. Chemical shifts may be quoted relative to CFCl$_3$ or to CF$_3$COOH and there maybe a few other standards in use as well for all we know. If you are following some literature data, always check which reference standard was used. In Table 11.1, we try to give a brief overview of ^{19}F chemical shifts in some of the more commonly encountered fluorine environments. The figures we quote are relative to CF$_3$Cl where the chemical shift of the fluorine is set at 0 ppm. Note that all shifts have negative values when using this standard.

Table 11.1 ^{19}F Chemical shifts of some typical fluorinated compounds (relative to CFCl$_3$ = 0 ppm).

11.9 Phosphorus (^{31}P) NMR

^{31}P is another nucleus which can be useful to the organic chemist. Many of the comments we have made about ^{19}F are also relevant to ^{31}P. Once again, compounds containing phosphorus are likely to show a clear ^{31}P 'signature' in their proton and carbon spectra and once again the chemical shift range for this nucleus is extremely large – as you might imagine, given the variable oxidation state of the element – typically −200 – +230 ppm.

The reference of choice for this nucleus is either H_3PO_4 or $(CH_3O)_3PO$. When H_3PO_4 is chosen, it is generally used as an external reference (80–85% solution) on account of its high acidity. The number and variety of possible organo-phosphorus compounds combined with the large chemical shift range over which the nucleus resonates make the inclusion of any useful NMR data somewhat problematic. We would refer you to more specialised texts or articles and also to some useful websites offering some enlightenment on the subject.

12

Dynamics

WARNING! This topic is complicated but really important! Please try to work your way through it because it is key to understanding what you observe in NMR spectra.

The British statistician, George Box, once stated that 'all models are wrong, but some are useful'. This is a valuable maxim to remember throughout your scientific career (or even beyond it!). With any model, there will be some simplification or approximation otherwise it won't be a model, but reality. This particularly shows itself in the chemical structure diagram (a.k.a. 'structure' or 'CSD').

Let's take an example structure (Structure 12.1).

Structure 12.1

Obviously, molecules don't look like this. It might 'look' a bit more like Figure 12.1.

But that doesn't really explain what we observe. In fact, we cannot really represent reality on a 2-D piece of paper. For example, a double bond describes a particular electron configuration between atoms. The hydroxyl (5) is shown to be connected to carbon (4) with a single bond, but it could also be possible that it is actually a double bond, with the bond between carbons (4) and (3) being single (keto-enol tautomerism). We use this simplified chemical structure diagram because it is quick to visualise and derive an understanding of the structural features of the

Essential Practical NMR for Organic Chemistry, Second Edition. S.A. Richards and J.C. Hollerton.
© 2023 John Wiley & Sons Ltd. Published 2023 by John Wiley & Sons Ltd.

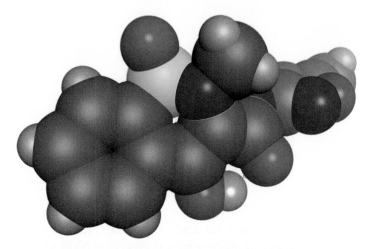

Figure 12.1   Space-fill molecular representation.

molecule. It is useful... but it is wrong. It is wrong because molecules are three-dimensional, and this diagram does nothing to represent three dimensions. Molecules may exist folded in solution where atoms which appear to be a long way from each other in the CSD are actually quite close.

So, what is this to do with dynamics? Well, in addition to the flaws in depiction of electron distribution, the CSD is also a static model of reality whereas molecules are constantly in motion, both with respect to each other and also internally. Single bonds in non-cyclic systems tend to have free rotation about them so, for example, all the protons on a methyl carbon appear equivalent. Rotation is fast on the NMR timescale (which we will come back to later) and so we observe the average environment that they experience. You can think of the molecule as a wobbly, rotatey thing, with all of the atoms moving about with respect to each other. If you have access to it, draw a structure in a molecular modelling package and perform a dynamics calculation and watch those atoms shake, rattle and roll! If you actually do this, you will be asked in the setup dialogue at what temperature you want to perform the simulation. Temperature is key. This is the energy available in the system to propel atoms into different environments. If there is a barrier to free rotation (for example, some steric interaction with a nearby atom), it can be breached by putting enough energy into the system to overcome that barrier. So perhaps a better representation might be (Figure 12.2):

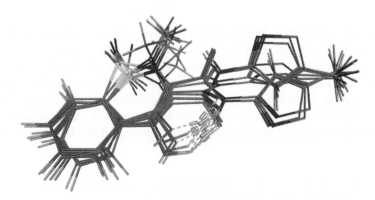

Figure 12.2   Different conformers.

Structure 12.2

At the risk of making this even more complicated, you need to remember that dynamics is statistical. At a given temperature, molecules will have a spread of energies, with the temperature representing their mean state. If there is a high energy barrier between two states, some molecules will overcome that barrier even if most of them won't. The higher the barrier is above the average energy of the system, the fewer molecules will find themselves in that environment.

If we take our friend, Structure 12.1, what does this mean for the keto-enol tautomerism? There is a complex answer to this. We have lots of things to consider. If we are in a protic solvent such as water or methanol, we might see one species predominate because of interactions with the exchangeable -OH proton. In an aprotic solvent it is quite likely that a different primary tautomer is observed. In this molecule, there is a carbonyl moiety (15 and 16) which looks like it is in a perfect position to form a hydrogen bond with the -OH if you draw the oxygen–proton bond explicitly (Structure 12.2).

A six-membered ring sets up a perfect angle for hydrogen bonding to occur so this may make the enol form most energetically favourable. Unfortunately, it still isn't as simple as that. The whole molecule shows conjugation. Amides often show double-bond character between the carbonyl carbon and the nitrogen (15 and 17 here). This may raise or lower energy barriers for specific rotations or proton locations for labile protons (those that move about). Tertiary amides often have such a high barrier to rotation that you will observe different species in the NMR timescale, leading to two sets of peaks. The unwary may think that the sample is a mixture of different compounds whereas it is a mixture of rotamers (the term we give to this phenomenon).

So, what are the consequences of this on interpretation of NMR data? Do we need to perform some sort of molecular modelling and dynamics calculation on everything? Well, fortunately not. There are many cases where we just need to bear in mind that dynamics exist. If you are looking at something simple like ethyl acetate, you see a triplet, a quartet and a singlet and that is fine. There is no need to think too much about dynamics (beyond the fact that all the protons on the methyl are equivalent due to free rotation of the carbon–carbon bond). Often, we start to think about dynamics more when we observe something in the NMR spectrum that doesn't fit a simple model, or when we see a structural moiety that we know is likely to misbehave (e.g. a tertiary amide).

So, let's take this example structure and run with it. Spectrum 12.1 shows the proton NMR spectrum of our molecule, run at 500 MHz and 30°C. The temperature is important, and we should really quote this every time we run an NMR spectrum. Often, we assume that the sample is run at room temperature but whose room and what temperature? In fact, many systems are set up to run at 30°C because this gives us greater temperature stability (heat the gas that

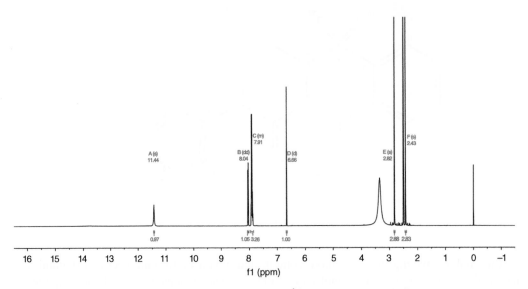

Spectrum 12.1   Proton spectrum of Structure 12.2 at 30°C.

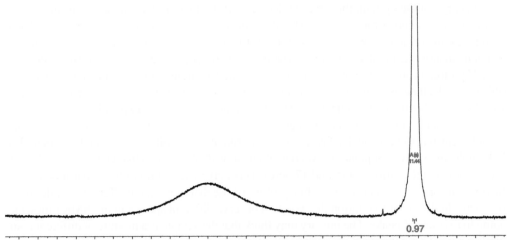

Spectrum 12.2   Expansion of low field region of Spectrum 12.1.

surrounds the sample tube and we can regulate this more reliably when it is higher than room temperature). We should, of course, always quote the NMR frequency because this will determine not only how we observe proton–proton splittings but also because it has an influence on dynamics. Once again, more later.

We have signals in the right places for most of the molecule. There are a couple of methyl signals at 2.44 and 2.83 ppm, for (21) and (1) respectively. We then have a one proton singlet at 6.67 ppm for the isoxazole proton (19). We then have a bunch of signals for the aromatic protons (7, 8 and 9 around 7.92 ppm and 10 at 8.05 ppm). This leaves our remaining two protons, the -NH and the -OH (if we have the enol form). There is only one proton visible at 11.45 ppm. If we boost up the vertical scale (a lot), we can just see another, really broad signal at around 13.6 ppm (Spectrum 12.2).

This leads to questions. Why is one signal sharp and the other broad? Which is which? The answer to the second question is: 'I'm not sure' although we would guess that the sharper signal at 11.45 is more likely to be the -NH (because it is analogous to many other aromatic amide signals that we have seen). The answer to the first question is, unsurprisingly, dynamics. The proton that we are looking at exists in multiple environments which have different chemical shifts. If the barrier were high between the states, then we would see multiple individual signals for each state. What we are seeing is a coalescence of the signals showing that the protons are experiencing different environments on the NMR timescale.

So, what is the NMR timescale? It is obviously important to this phenomenon. We define the NMR timescale as the length of time that it takes for an excited atom to return to its lower energy state (normally the time for a signal transient). For a 1-D proton experiment, this is a few seconds. Going back to the chapter on theory, if we pulse and then acquired an FID, we will be observing the behaviour of protons over that time. If a proton changes its behaviour whilst we are observing it, we will see the average of its behaviour. This is why our NMR timescale is defined in this way. Dynamic behaviour is relevant to many analytical observations although the timescale is different. HPLC, for example, is likely to have a timescale measured in minutes, so we are unlikely to observe some of the phenomena that we do in NMR as they will be averaged to a single value.

Dynamics has an impact on all that we observe in NMR. Linewidths, chemical shifts, splittings and relaxation pathways are all significantly influenced.

12.1   Linewidths

The natural linewidth of an NMR signal is very narrow. Each signal is made up of a number of transitions. They are merged into the observed (and simplified) signal through several dynamic effects. There will be geometric changes in the molecule, tumbling in solution and spin–spin and spin–lattice interactions. What we are observing is a composite of all of these things. Frightening, isn't it? However, it doesn't need to be. We need to be aware that all of this stuff is happening, but we don't have to do a full *ab initio* analysis for every spectrum that we look at. We can get most of the information that we need by looking at the gross features of the spectrum. Where we start looking deeper is that case when the alarm bells start ringing and make us question whether what we are observing fits our proposed structure.

12.2   Chemical Shifts

The observed chemical shift (as we have seen earlier) is a result of the weighted distribution of all conformational forms in solution. As we cover in Chapter 15, this is one of the reasons that proton NMR prediction is so difficult. The observed chemical shift is directly influenced by the effective magnetic field that the particular nucleus experiences. This, in turn, is impacted by not only intramolecular interactions, but also intermolecular interactions (for example, with the solvent). Because this is a dynamic phenomenon, it is affected by temperature and may also be affected by concentration and pH. We can expect chemical shifts to be exactly the same, only if the experimental observations are performed under identical conditions. This is really hard to achieve so you can always expect chemical shifts to vary from experiment to experiment.

12.3   Splittings

Chapter 1 explains where splittings come from and Chapter 6 talks about the geometric relationship controlling coupling constants (which we observe as splittings). Since the geometric relationship between nuclei is determined by the conformation of the molecule, it is obvious that this will be strongly influenced by dynamics. We are observing an averaged view of the environment that the nuclei observe.

12.4   Relaxation Pathways

When we perform NOE experiments, we are trying to identify spatial relationships between nuclei. The NOE effect has an r^6 dependency on distance(r) so that only close nuclei will interact. The normal threshold we expect (without degassing or other techniques to maximise the NOE) is an observed effect when the nuclei are within about 4 Å of each other. If our molecule of interest has a small population of a conformer where the two nuclei come very close, because of the r^6 term, we may still see a significant NOE even though the majority of conformers would not show an enhancement. This means that you need to be very careful when looking at NOE data and bear in mind that your observation may be influenced by a minor conformational population.

12.5   Experimental Techniques

How do we know what we are observing is a dynamic effect and not just a mixture of compounds? An easy experiment is to perturb the equilibrium. If we heat or cool the sample, we will force the equilibrium to a new position. If we observe two sets of signals, we can heat the sample to try to speed up the interconversion between the different states. There is a limit to how high a temperature we can heat the sample. It may decompose but also the probe will have a maximum temperature at which it can operate. This is often around 120°C although some probes may be lower than this or may become difficult to shim as they increase in temperature. You need to check with your particular probe to know what you can get away with. Of course, you may heat the sample but be unable to coalesce the signals because the barrier is too high. If you have seen some degree of coalescence, this may be enough to confirm that the observation is due to rotamers. If, on the other hand, you need to coalesce the signals to help interpret the signals then you might be out of luck. If you are running the samples at a high field, e.g. 600 MHz or more, you can try running them at a lower field as this will decrease the barrier slightly.

We haven't mentioned cooling the sample and this is because it almost always doesn't help. If you have broad signals that you want to make into sharp signals due to dynamics, in theory you could freeze out the forms by cooling the sample. You have a limit of how low you can go. If you are in DMSO, forget it because DMSO freezes at 19°C! If you are in chloroform, you can go down to −63°C but you will find your solution getting more viscous and the tumbling of molecules in solution will be slowed down, making the spectrum really broad. You may also have a low temperature limit on your probe too. There are probably some cases where this makes sense but not very often in our experience.

12.6 In Practice

Let's look at a real example of a sample that is affected by partial double-bond character of a tertiary amide (this is something that happens a lot). If we acquire the data at 30°C we end up with this spectrum (Spectrum 12.3):

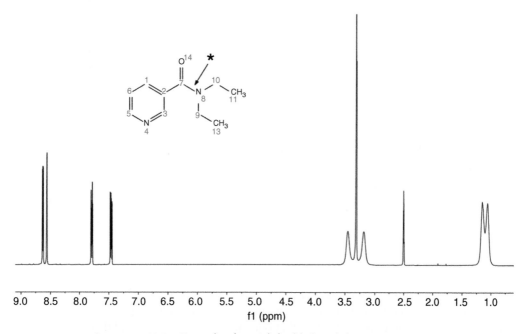

Spectrum 12.3   Example of partial double bond character at 30°C.

Normally, we would expect the two ethyl groups to be equivalent but, in this case, we are seeing two broad signals for the methyl (~1.1 ppm) group and also for the methylene (~3.3 ppm). The signals for the pyridine are sharp and exactly as we would expect. What we are observing is the lack of free rotation around the bond between the carbonyl carbon and the nitrogen. This is causing the two ethyl groups to be non-equivalent because one will experience an environment where it is close to the pyridine and the other will be closer to the carbonyl. If you picture the bond between 7 and 8 being a double bond, then you would expect the two ethyl groups to be different, although they would be sharp and well-defined, unlike this example. The broadness is telling us that there is some dynamic interconversion taking place which is what we would expect if that 7–8 bond has partial double-bond character. Note that we are seeing a 50:50 distribution between the two rotamers. This is because we have the same groups attached to the nitrogen. If we had two different groups, then the proportions of the rotamers could be very different. The fact that the groups are the same is also the reason that the pyridine signals aren't affected by the rotamers – they will see the same environment whichever ethyl is next to them. If we had two different groups, especially if one had a high degree of anisotropy (e.g. a phenyl group or a carbonyl) then we would expect to see doubling-up of the pyridine signals too.

In this simple case, we would probably look at the spectrum and move on because this is exactly what we would expect. In more complex cases, especially when the populations of the rotamers are very different, we might want to convince ourselves that smaller peaks are not impurities. So, let's heat this one up to 120°C to see what happens (Spectrum 12.4):

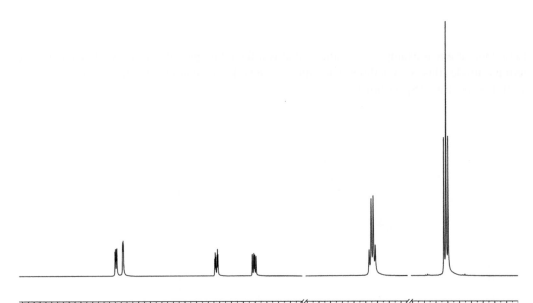

Spectrum 12.4   Example of partial double bond character at 120°C.

Our broad lumps have now coalesced into a recognisable triplet and quartet as expected. Those of you with a sharp eye will notice that our peak-shape isn't that good. There is a slight tailing to the right of the peaks indicating that our shimming isn't great and that z^2 is probably too low (see Section 3.9). This spectrum was acquired under automation, and we can struggle to shim hot samples for a number of reasons. The heat may cause a subtle change in the geometry of the probe components but, more likely, is that the heat is causing convection in the sample, making it non-uniform. We can minimise the impact of this by not over-filling the NMR tube. The longer the column of solvent, the more likely that different parts will be at different temperatures, setting up convection. Obviously, we need enough to fill the probe coils (see Chapter 2 on sample preparation) so we can only reduce the solvent column by a certain amount.

One last thing to watch out for is that the chemical shift of the signals may change. This is the chemical shift of all the signals, including TMS and the solvent signal. In this case, we have referenced to TMS and can see that the DMSO signal moved from 2.47 ppm to 2.44 ppm. The water signal moves from its normal ~3.3 ppm to about 2.8 ppm. Incidentally, this can be a way to move the water signal out of the way if it sits on top of a signal that you are interested in. This trick can be useful if you need to quantify a signal where the water interferes. Whilst we are talking about quantifying, be aware that it becomes easier to saturate signals at higher temperature because the extra energy will promote more signals into the higher energy level. If you are working at high temperature and quantification is important, make sure to have a long relaxation delay.

12.7 In Conclusion

Dynamic effects may cause confusion when they exhibit themselves in your spectrum. If you observe something that you don't understand and think that it might be a dynamic effect, you need to perturb the environment to move the equilibrium. Perturbation is normally achieved thermally but you can also push some equilibria by changing pH.

If you are using heat and the barrier is high, you may not be able to coalesce the signals but you may see them start to move together or broaden and this will give you more confidence that your attribution of the observation to dynamics is correct. If you cannot coalesce the signals and heating is inconclusive, you can also irradiate one of the signals with a decoupler and this will effectively irradiate its partner if they are interconverting. The easiest way to see this is to use a 1-D NOE experiment and your irradiated signal and its partner will be shown as negative peaks if they are interconverting.

13

Quantification

13.1 Introduction

NMR offers us a great tool for quantification. This is because it inherently has a uniform response to the nucleus of interest (see caveats at the end of this chapter). We rely on this when we look at integrals in a proton NMR spectrum – a methyl group integrates for three protons, a methylene integrates for two protons, etc. As NMR spectroscopists, we get a little blasé about this – we just expect it. This is not true for all techniques though. For example, ultra-violet (UV) detection is often used on HPLC systems but its response depends on the degree of conjugation in the compound of interest. If we were to have a chromatogram with two different compounds in it, we would not be able to tell what their relative proportions were unless we knew their UV response at the wavelength (or wavelengths) being monitored. In NMR, this is not the case…

13.2 Different Approaches to Quantification

Depending on what you are after, there are many different ways of using NMR to quantify materials. Some of these are fairly simple but others require considerable thought to achieve suitable quantification. The accuracy that you require will determine the effort that you have to put into quantification. If you want to get to about 20% accuracy, then life is easy but as you try to get more and more accurate, the amount of effort and thought goes up considerably! Here are some of the different approaches to quantification.

13.2.1 Relative Quantification

This is the easiest case for NMR (and other analytical techniques). What we are looking for is the relative proportion of compounds in a mixture. To do this, we identify a signal in one compound and a signal in the other. We then normalise these signals for the number of protons that they represent and perform a simple ratio calculation. This gives us the *molar* ratio of the two compounds. If we know the structure (or the molecular weight) of these compounds, then we can calculate their *mass* ratio.

Essential Practical NMR for Organic Chemistry, Second Edition. S.A. Richards and J.C. Hollerton.
© 2023 John Wiley & Sons Ltd. Published 2023 by John Wiley & Sons Ltd.

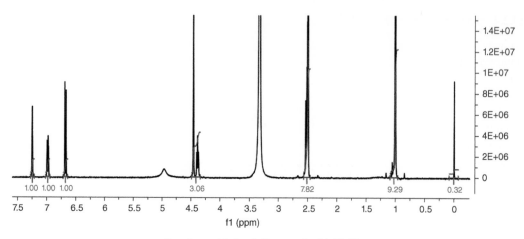

Spectrum 13.1   Salbutamol with TMS.

Spectrum 13.1 shows a spectrum of salbutamol in D_6-DMSO with some TMS in it. As an exercise, we can easily quantify the TMS as follows…

The signal for TMS (0 ppm) is for 12 protons. The signals in the aromatic region are from salbutamol and represent one proton each. If we set the integral of the aromatic protons to equal 1.0 and assuming adequate relaxation time for the relevant protons of both salbutamol and TMS, then we see that the relative integral of the TMS is 0.32. Because this signal is for 12 protons, we can calculate that we have $(0.32/12) \times 100 = 2.6$ mol% of TMS in the sample. The molecular weight of salbutamol is 239 and the molecular weight of TMS is 88 so their weight ratio is 0.36 which means that the weight ratio of TMS is $2.6 \times 0.36 = 0.96\%$w/w.

13.2.2   Absolute Quantification

The previous example is fine if all you want to know is the relative proportions of compounds in your solution. If you know the *absolute* concentration of one of the components, then you can work out the absolute concentration of the other as a result.

13.2.3   Internal Standards

If we add a known amount of a compound to our solution, we can use it to quantify the material of interest. This is great except that we may not want to contaminate our material with some other compound. A number of people have looked at using standards that are volatile so that they can be got rid of later (TMS is an example that we have seen published). The problem with this approach is that if the standard is volatile then you need to run it quickly before it disappears. TMS disappears really quickly from DMSO so it is probably not a good idea in this case. TMS also suffers from the fact that it has a long relaxation time so you have to be very careful with your experiment to ensure that you do not saturate the signal. The last major problem with TMS is that it comes at the same part of the spectrum as silicon grease which can be present in samples. Choosing a standard so that it has a short relaxation time, is volatile and comes in a part of the spectrum free of interference is really tricky. In fact, we wouldn't recommend it at all.

13.2.4   External Standards

So how do we quantify if we don't have an internal standard? One way is to use an external standard. This is done by inserting a capillary containing the standard into the NMR tube (Figure 13.1).

Of course, we still have the problem of selecting a compound that doesn't interfere with the spectrum and that has a suitable relaxation time but we don't need to worry about its volatility. What would be really good is a standard that doesn't interfere with the sample at all. Something that has no relaxation time to worry about and something that you could put in the spectrum in an area where there were definitely no signals...

Figure 13.1   External standard.

13.2.5   Electronic Reference (ERETIC)

This problem was eventually solved in the magnetic resonance imaging world where they needed to be able to quantify things *in vivo*. The result was the use of an extra radio-frequency source during acquisition. It was called 'ERETIC' (**E**lectronic **RE**ference **T**o access *In vivo* **C**oncentrations) and has been used extensively in recent years in the high-resolution liquid NMR areas. The great advantage of this approach is that you can make the signal as big or as small as you want and put it anywhere in the spectrum (−1.0 ppm is a favourite place). The way that you use it is to quantify the ERETIC signal against a sample of known concentration. Once this has been done, you can then insert the signal into your unknown concentration spectrum and integrate it against one of the signals in your compound.

There are some problems with the ERETIC approach. Firstly, it does not respond in the same way as the signals in your sample do so if your probe tuning is not quite right, you will get an inaccurate answer. Secondly, it requires rewiring of your system so that you can introduce the signal (alternatively, you can rely on crosstalk in the system to let the signal bleed through – this too has some problems associated with it). Lastly, because the signal is generated in a different manner from those of the sample, it can suffer from phase-errors which give rise to inaccuracy when integrating the signal.

13.2.6   QUANTAS

Given that the ERETIC approach has problems, why not introduce a defined intensity signal into your spectrum using software? This is the approach adopted in the QUANTAS technique (**QUAN**tification **T**hrough an **A**rtificial **S**ignal). No, not the Australian airline (which is 'Qantas', by the way). In this approach, a reference spectrum with a single signal is created using software. This is added to a spectrum of a sample of known concentration and a scaling factor is calculated to make the signal exactly the correct size for the concentration that it is to represent. If this spectrum is added to any sample spectrum, using the calculated scaling factor, it will be able to represent a defined concentration in the unknown concentration spectrum. This is shown in the salbutamol spectrum used before (Spectrum 13.2).

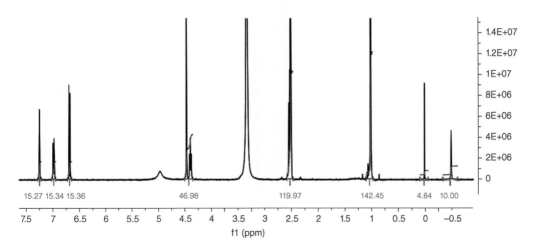

Spectrum 13.2   Spectrum with QUANTAS signal.

Unlike ERETIC, this approach does not track receiver gain or number of scans (the signal is a fixed intensity). This doesn't cause a problem though – you can choose to run under identical conditions to your reference, or you can compensate for differences in acquisition condition. For example, the signal builds directly proportionally to the number of scans (note, not the square root of the number of scans. It is the signal-to-noise that builds with the square root of the number of scans). On modern spectrometers, receivers are linear and it is possible to compensate for receiver gain differences linearly too. The current implementation of the QUANTAS method uses a small program to automatically take into account any changes in receiver gain and number of scans so you just end up with the signal at the correct level.

This approach offers by far the most simple and flexible way of quantifying samples and is even better because it can be run retrospectively on any sample (as long as the spectrometer is performing similarly to when the signal was standardised). It turns out that for most modern spectrometers, the spectrometer is stable over many months or even years.

13.2.7   ERETIC2

Not to be confused with ERETIC! Bruker unfortunately used this term for their equivalent of the QUANTAS approach (both of which are based on PULCON by Wider & Dreier). ERETIC2, like QUANTAS, has been wrapped in a user-friendly piece of software to enable simple quantification.

13.3 Things to Watch Out For

It all seems so simple when you look at this example. Unfortunately this is not necessarily the case. We need to be a little careful about how we acquire the data if we are going to use it for quantification.

The first thing to look out for is the relaxation time (T_1) of the protons that you are going to measure. In order to get an accurate integral, the protons must return to their rest state each time before you pulse them. The recommendation for a 90° pulse is to wait for 3–5 times T_1. Obviously this assumes that you know the T_1 of all of your protons. It is possible to measure them (and this is indeed the 'right' thing to do) but you need to decide how accurate you need the result to be. If you want a fairly accurate result, it is sufficient to 'guesstimate' your T_1s just by looking at the chemical structure. Small molecules tend to have long T_1s. Methyl groups tend to have longer T_1s than methylenes. Methines may have long T_1s if they are isolated from any other protons. Symmetrical molecules have slightly longer T_1s than unsymmetrical molecules. If you use a 30° pulse (which is more normal) then you can probably get away with using a relaxation delay of about 5 seconds if your acquisition time is about 3 seconds (hence a total recycle time of about 8 seconds).

On older spectrometers, it is important that the signal that you are measuring is not at the edge of the spectrum. This is because older spectrometers used hardware frequency filters and these start to decrease signal intensity at the edge of the spectrum. More modern spectrometers use digital filters that are capable of very sharp cut-offs that will not affect the intensity at the edge of the spectrum. Be warned, even here you may get problems with distortions in the baseline at the edge of the spectrum (the so-called smileys). In general, try to avoid your signal of interest being at the edge of the spectrum.

All quantification relies on being able to standardise against a known concentration standard. This is not a trivial thing to do as it requires an accurately weighed amount of a known purity compound, made up accurately to a precise volume. If your standard is wrong, all your measurements will be wrong so it is worth spending some time getting it right!

Ultimately, you will be measuring and comparing integrals so you need to be very careful about how you get these. Your signals of interest must be perfectly phased, clear of other signals and on a good baseline. The integrals must also have good slope and bias (which they should do if everything else is correct). Any problems with any of these variables will seriously degrade the accuracy of your result. In our experience, the biggest single error with any NMR quantification approach is the error in measuring the integral. If you need really accurate integrals, it may be better to use peak-fitting routines to calculate the peak areas. Most processing software has this capability and it can take some of the subjective nature of integration out of the process (Figure 13.2).

13.4 Quantification of Other Nuclei

You can use this approach to quantify other nuclei – it works just as well for [19]F. It doesn't work very well for ^{13}C because we normally acquire ^{13}C data with proton decoupling (which creates an NOE enhancement) and we acquire it far too quickly so the signals are not quantitative. If

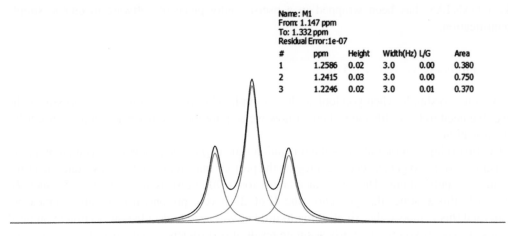

Figure 13.2   Performing peak-fitting.

we have lots of material, we can lose the NOE enhancement by decoupling only during acquisition. We can also use a trick to make the carbon atoms relax more quickly, such as adding a paramagnetic relaxation material called Cr(acac)3 (or 'Crack-ac' as it is known). So, whilst it is possible to quantify ^{13}C, it is something that we would normally avoid and look for an alternative.

13.5   Conclusion

If you do manage to get everything right, NMR offers excellent quantification results. What's more is that it is free if you have acquired a 1-D spectrum. The amount of work involved in quantification is related to the level of accuracy required. For ultra-high accuracy, significant effort should be expended on every aspect of the analysis and the protocol should be validated with known amounts of material. If you are after ±10% accuracy, then there is almost no effort involved, making NMR a first choice for quantification in our opinion!

14

Safety

NMR systems are pretty safe if treated correctly but this short chapter outlines some of the things you may need to think about when using them. Note that we are not pretending to offer a full safety assessment but this should alert you to the major hazards associated with modern NMR systems. There are very good documents available from the major NMR manufacturers which cover this area in considerable detail.

NMR-associated risks fall into three categories: Magnetic Fields, Cryogens and Sample-related Risks. If you were to be picky, then you could also add things like risk from electrical shock, etc. but we're sure that you wouldn't be.

14.1 Magnetic Fields

It might seem obvious but magnetic fields can attract magnetic materials towards them. It is something that we are all used to but most people haven't experienced the strength of field from an NMR magnet. Modern magnets are often shielded but they still have quite strong external fields, especially at the base of the magnet. Because the field is invisible it is easy to forget that it is there. We know of a case where a photographer forgot about the field and moved his tripod closer to the magnet to get a better shot. The next thing he knew, his tripod flew across the room and smashed into the magnet, damaging it in the process. It took nine months to get the magnet back to field and a lot longer to live down the embarrassment. Had someone been between the tripod and the magnet it could have been even more serious. There is also a story (maybe an urban myth?) about someone who had a metal ruler in their lab coat pocket while he did some work at the base of the magnet. The ruler flew out and sliced the end of his nose off.

As well as the obvious risks, there can be less obvious risks too. Heart pacemakers can be disrupted by strong magnetic fields so this needs to be pointed out to anyone who enters the area in case they are reliant on one. Another risk is for people who have certain metal prosthetics (e.g. hip joints) – you wouldn't want them stuck to the side of the magnet, would you? Another example that we have had is with metal breathing apparatus – someone was pulled back to the magnet when wearing it during a fire drill.

Essential Practical NMR for Organic Chemistry, Second Edition. S.A. Richards and J.C. Hollerton.
© 2023 John Wiley & Sons Ltd. Published 2023 by John Wiley & Sons Ltd.

There can be other untoward effects which may not be exactly safety-related but are inconvenient to say the least. Magnetic strips on credit cards can be erased, hard disks can become corrupted (watch older iPods near a magnet) and electrical equipment with relays can be affected. Additionally, strong fields may saturate transformers so that they don't behave as expected – probably not a good thing if you were expecting a particular voltage out of them. Lastly, analogue watches fare badly in strong magnetic fields. If you are lucky, the watch may just lose time while it is close to the magnet. If you are unlucky, it may stop and refuse to start again. As a rule of thumb, don't get close to the magnet and wear a watch unless it doesn't have hands.

So how close is too close? Well, it depends on the magnet, whether it is shielded or unshielded or if it is a wide-bore or normal magnet. The magnet manufacturer will be able to give you the figures. We measure the safe distance by the stray field at that point. Normally we take the 5 Gauss field line as a safe distance and this is normally marked in some way (tape on the floor or a physical barrier). For a 400 MHz shielded spectrometer, this distance is about 1–1.5 meters from the centre of the magnet. The larger the magnet, the further the stray field (generally), although modern shielded magnets may have the 5 Gauss line at the edge of the magnet can. Figure 14.1 shows the 5 Gauss line for a set of magnets.

There are no known adverse biological effects of static magnetic fields and it has been deemed that they do not cause problems to people working with them. This is always under review and there may be moves to limit time spent in strong magnetic fields at some time in the future. Moving quickly in a strong magnetic field may cause disorientation due to pushing the fluid that circulates in the inner ear. This isn't too much of a problem in an NMR lab but may be more problematic in an MRI scanner, where the head is exposed to a much stronger field.

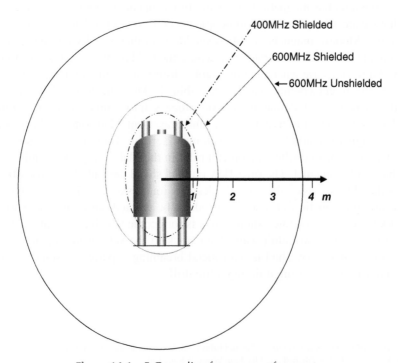

Figure 14.1   5 Gauss line for a range of magnets.

14.2   Cryogens

Because the magnet is essentially a Dewar containing liquid nitrogen and liquid helium, the biggest risk is from those cryogens. The first risk is from the low temperatures of these liquids (liquid helium boils at about −270°C) which can cause serious burns. The second risk is of asphyxiation as the cryogens boil off.

The risk of burns is normally only experienced when filling a magnet with nitrogen or helium. You need to be protected in case the liquid spills or the transfer line breaks. Protection just means covering up any exposed skin (lab coat, visor and thick gloves are normally sufficient). At all other times, the cryogens are safely in their cans and should stay there unless something catastrophic happens.

Asphyxiation is another invisible hazard. If the oxygen level in a room decreases below a certain level you may become unconscious and die. In general, big labs are better than small labs because the natural volume of the room will help to dilute these effects. Likewise, an efficiently air-conditioned room will change the air in the room fairly rapidly and this will also help keep oxygen levels up. If your NMR magnet is in a small room, it may be necessary to install oxygen depletion sensors. These will alert you should the oxygen level fall below a safe value.

When a magnet is not being filled, it will give off a steady stream of nitrogen and helium. The helium will normally sit near the ceiling whereas the nitrogen will tend to permeate the whole volume of the room. When your instrument is installed, a survey should have been carried out to evaluate the risks.

Lastly (on the subject of asphyxiation) when transporting Dewars of cryogens it is important that you and the Dewar are not in a small space together. This includes lifts (elevators). If you need to transport a Dewar up or down floors in a building, you should send the Dewar on its own and prevent people from joining it!

There is a special condition that can arise in an NMR magnet, called a 'quench'. This occurs if the magnet coils suddenly cease to be superconducting and all the energy stored within them is released as heat. This causes the helium in the can to boil off very rapidly. There are two major risks from this. The first is obviously asphyxiation; the second is the pressure that is generated by the increased gas volume. To minimise problems from the latter, the room should be constructed so that the gas can escape quickly. The other precaution is to ensure that the doors to the lab open outwards, otherwise the gas pressure may make it difficult to open them, trapping the occupants. The asphyxiation risk from a quench is quite low because the helium has a tendency to sit at the ceiling of the room and it also escapes very rapidly from wherever it can. Nonetheless, it is advisable to leave the room if a quench happens. You will know when a system is quenching – it makes a lot of noise and you get clouds forming in the ceiling!

The event that would cause the largest release of cryogens would be a catastrophic failure of the can. This would release the helium and the nitrogen very quickly. Fortunately, this is an unusual event and previously mentioned precautions should still work.

One last hazard with cryogens is that they may lead to local build-up of oxygen through liquefaction of air. When filling a magnet, it is possible to see liquid air condensing at the fill port. If this happens, there is a risk of causing combustion of oil or other materials that are close to the liquefied air. This risk can be eliminated by keeping sources of combustion away from the magnet and Dewar.

14.3   Sample-Related Injuries

While the potential hazards associated with powerful magnetic fields and cryogens are spectacular, it's the everyday hazards associated with the handling of NMR samples that are most likely to catch out the unwary! Standard 5 mm NMR tubes are very fragile (3 mm even more so) and the thin-walled glass tube they are made from can cause nasty cuts. Pushing on the plastic tube tops is the most dangerous part of the process as considerable force is sometimes needed. We have found that it is safest to hold the tube in one hand and lay the knuckles of that same hand into the palm of the other which is used to push on the top. Locking your hands together in this way, minimises the chance of injury should the tube shatter, as there will be no danger of sudden violent movement of flesh towards broken glass. Having worked in NMR for a long time, we have learned the lesson through experience. This hint will save you from having the same experience!

Another source of danger relates to the samples themselves. In a research environment, many of the compounds made are of totally unknown toxicity and so should be handled with extreme caution. NMR solvents are obviously toxic in their own right but when they contain unknown organic compounds in solution, the hazards are far worse. Be warned that all organic solvents commonly used for NMR can pass through skin and into the bloodstream but DMSO is particularly good at it. If this happens, it will take anything dissolved in it through as well so avoid spilling any solutions on your hands while making up or filtering samples. Don't be lulled into a false sense of security by wearing thin rubber gloves – they offer little protection as solvents can penetrate them too.

15

Software

There are many software tools available to help with the acquisition, processing and interpretation of NMR data. Attempts have been made to automate the verification process and even perform full structural elucidations of unknown compounds. As you might guess from the complexity of the interpretation chapters, these software solutions are not foolproof! It remains to be seen whether they ever will be good enough but there have certainly been some major steps forward in all of these areas.

In this chapter, we will look at the different types of software, but be warned that software development is quite dynamic and the landscape may be very different when you come to read this section! However, when producing the second edition of this book, it was surprising that things have not moved on as much as might be expected in the software area. Some of the smaller (in terms of usage) applications have disappeared but the big players remain the instrument vendors (as might be expected) along with ACD/Labs and Mestrelab. There have been significant improvements in assisting spectral assignment with both ACD/Labs and Mestrelab offering tools to help indicate likely atoms in the chemical structure while moving over multiplets in the NMR data (and the other way around).

15.1   Acquisition Software

You seldom have much choice about this software. When you buy a spectrometer you will get some software from the manufacturer. The main manufacturers are Bruker, Varian/Agilent and JEOL. Their software is called: Topspin, VnmrJ and Delta respectively. Since the first edition of the book, Varian were acquired by Agilent who subsequently withdrew from the NMR sector. There are still many Varian/Agilent spectrometers in the wild so VnmrJ is still a current application. The VnmrJ software was made open source and is now called 'OpenVnmrJ'. Note that OpenVnmrJ is not available for Windows PCs but only for certain Linux distributions and MacOS.

These pieces of software are quite complex as they have to perform all the spectrometer control as well as processing and some simulation. That said, all manufacturers have improved their software to make it more user-friendly in recent times and it is not the challenging beast that it used to be.

Essential Practical NMR for Organic Chemistry, Second Edition. S.A. Richards and J.C. Hollerton.
© 2023 John Wiley & Sons Ltd. Published 2023 by John Wiley & Sons Ltd.

The advent of benchtop NMR systems have seen a new generation of acquisition software targeted at each vendor's systems. In most cases, data processing is carried out using an existing commercial package. Some companies have specialised in taking old CW instruments and converting them to FT. They too tend to use commercial NMR processing software.

15.2   Processing Software

As mentioned above, the manufacturers provide software to process the data. These pieces of software are designed to process data created on that manufacturer's instrument although they can process most data from other vendors (sometimes this is not as easy as it could be). In addition to manufacturers' software, there are also third-party software suppliers who offer software capable of processing data from all makes of NMR spectrometer. At the time of writing, there are a number of these companies; the most well-known of these are probably Advanced Chemistry Development (ACD/Labs – http://www.acdlabs.com) and Mestrelab Research (http://www.mestrelab.com). ACD's product for processing is called ACD/Spectrus, Mestrelab's product is called MestreNova (or Mnova for short).

Figure 15.1 shows a screenshot of one example of processing software. There are many other pieces of processing software out there, some good, some not so good. If you don't have any and you need some, it is worth having a look around to see what is available. In our experience, you tend to get what you pay for and the more expensive software is generally better. On the other hand, one of the cheaper packages may do just what you want in which case you have a bargain! Here is a list of some of the more well-known packages:

Name	Website	Platform
ACD/Spectrus	acdlabs.com	PC, Web
Mnova	mestrelab.com	Mac, PC, Linux, Tablets
iNMR	inmr.net	Mac, PC
NMRPipe	www.ibbr.umd.edu/nmrpipe/	Unix, Linux
FELIX	www.felixnmr.com	PC, Linux, Unix
JASON	www.jeoljason.com	PC, Mac
NMRnotebook	www.nmrtec.com	Mac, PC, Linux

If you search the internet, you will find many other NMR processing packages available. Some of them are specifically designed for a particular use of NMR, e.g. proteins or metabolomics and so outside the scope of this book. If you are an academic or student, you may find that some of these software packages are free (or a reduced price). It is always worth getting in touch with the company to find out.

Basic processing will allow Fourier transform, phase correction, baseline correction, integration and output to a printer/PDF/clipboard. More advanced packages may perform multiplet analysis, peak-fitting and assignment to a chemical structure. This can speed up your interpretation considerably. Even better, they may cascade your assignments through a set of 1-D and 2-D spectra. This capability is available in Mnova and Spectrus Workbook. Note that you can't do this in Spectrus Processor (the entry-level application from ACD/Labs). You may need to watch out for how Non-Uniform-Sampled data (NUS) is handled as processing this data requires special algorithms.

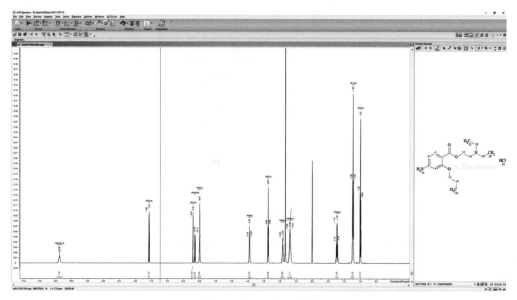

Figure 15.1 Example of NMR processing software.

15.3 Prediction and Simulation Software

You will normally have access to the previous two categories of software as a minimum – they will be on the spectrometer itself. One thing that you don't get supplied is software to predict chemical shifts (although you may get some sort of simulation software). The desire to predict the chemical shift of a nucleus has been around since the first time that the chemical shift phenomenon was discovered. There are numerous papers going back to the earliest days of NMR trying to relate structural properties to chemical shift. Early work was concentrated on proton chemical shift prediction (because carbon data was so hard to get) but it was soon realised that the unpredictable nature of proton chemical shifts (their dependency on average solution conformation) made this job difficult. It was easy to get to within about 0.5 ppm of the correct shift but this is not too good when 90% of your chemical shifts come within a 6 ppm range. Apart from generating additivity tables (as used in this book), proton chemical shift prediction was soon ignored.

15.3.1 ¹³C Prediction

Once ¹³C data was more readily available (with the advent of FT spectrometers), interest in chemical shift prediction was re-born. The reasons for this were that carbon spectra don't show carbon–carbon coupling information (unlike proton–proton coupling) and so knowledge of carbon chemical shifts was really important in the assignment of ¹³C spectra. There were numerous efforts at carbon prediction but perhaps the first truly successful method was created by Wolfgang Bremser in 1977. He realised that you needed to have a standard way of naming and sorting carbon atoms so that you could look them up in a table. He also realised that if you were methodical about this and used a naming system that grew in 'shells' (Figure 15.2) from the atom of interest (atoms closest to the atom of interest come first in the name), you could predict the chemical shift of a similar compound by interpolating between entries in his table. He named this the HOSE code (Hierarchically Ordered Spherical description of Environment).

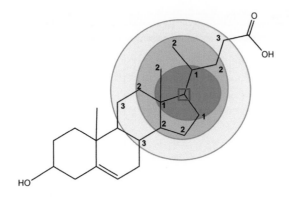

Figure 15.2 HOSE code in operation.

Bremser produced tables of these HOSE codes from NMR work that was carried out at BASF in Germany. Most modern carbon prediction routines still use this HOSE code today (albeit slightly modified from the original). Modern software hides all the HOSE code generation in the background so all you do is draw a structure and press the predict button and all the chemical shifts are calculated.

Modern carbon prediction software has hundreds of thousands of chemical structures to call on (Bremser had about ten thousand when he started). The more structures you have, the better the chance that something similar to your structure will be in the database – and the better the quality of the chemical shift prediction. On the other hand, you may have something that has a fragment that just isn't represented in the database – in which case you cannot predict the chemical shift accurately.

There are a number of ^{13}C prediction packages that are commercially available. These are often embedded in packages such as ACD Spectrus or Mestrelab Mnova (Figure 15.3). You can also find simple additivity approaches in chemical structure drawing packages such as ChemDraw.

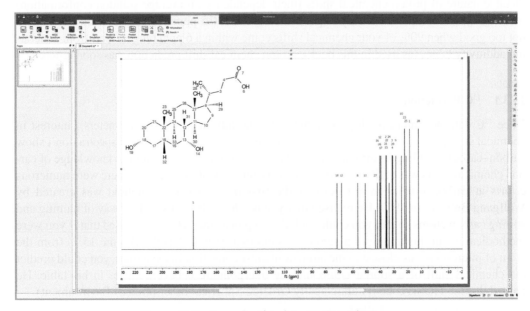

Figure 15.3 Example of carbon NMR prediction.

All in all, carbon prediction is really very good. This is partially due to the huge amount of carbon data in the public domain, partially because carbon chemical shifts are distributed over 200 ppm (instead of 10 ppm for proton) and partially because carbon chemical shift is mainly influenced by through-bond contributions rather than through space (hence less dependence on conformation). Nonetheless, you need to be a little discriminating when assessing the chemical shift values that these systems come up with – if you have a carbon atom in an unusual environment and this is not covered in the database, the prediction bears very little weight and you must rely on other information to assign the atom.

15.3.2 ¹H Prediction

Despite having been the earliest attempted prediction, proton prediction remains relatively poor. The reasons for this have been alluded to earlier but to summarise; the proton chemical shift is often highly dependent on through-space effects (anisotropy) and has a very small distribution. Figure 15.4 shows the observed intensity-weighted distribution of peaks harvested from a random selection of 40,000 spectra (the large grey bars are solvent signals). You should always treat a proton prediction as a suggestion rather than the truth! In particular, prediction of exchangeable proton chemical shifts is very difficult as they are strongly affected by pH, concentration and temperature.

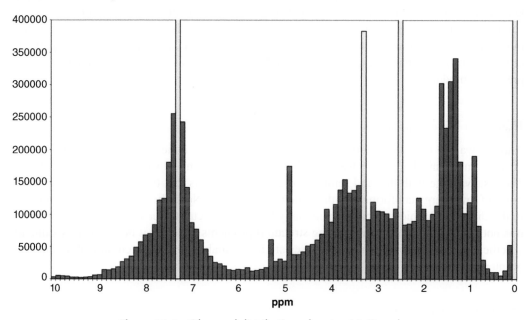

Figure 15.4 Observed distribution of proton NMR peaks.

Whether performing proton or carbon prediction, there are five main commercial approaches to proton prediction currently: Incremental parameters, HOSE code databases, semi-empirical, *ab initio* and neural network methods.

15.3.3 Incremental Approaches

These are computerised versions of the tables in this book. Chemical shifts are calculated by adding together the contributions from the various functional groups attached to the core of interest. These are normally split into three types: aromatic, aliphatic and olefinic. Probably the best example of this approach is the Upsol predictor that was supplied with the book *Structure Determination of Organic Compounds* by Pretch et al. (Springer). This has found its way as an add-in to a few commercial systems. The advantage of this approach is that calculations are quick and it is very easy to implement (and hence low cost). The disadvantage is that it is not very accurate and becomes progressively less accurate as more substituents are added to a core.

15.3.4   HOSE Code Databases

ACD/Labs and Mestrelab have extensive databases which use this approach. Predictions are generally very good for carbon prediction but less so for protons. Anisotropy (through space interactions) are not modelled and this can have a very large effect on proton chemical shifts, even for structural fragments that are well-represented in the database.

15.3.5   Semi-Empirical Approaches

Currently (to the best of our knowledge) there is only one product that adopts this approach and this is NMRPredict from Modgraph. It is based on the work by Prof. Ray Abraham at the University of Liverpool. This approach calculates chemical shifts for a range of low energy conformers and averages them to give a net chemical shift. This approach seems to offer the most accurate prediction of chemical shift but the disadvantage is that it is very slow (particularly for conformationally flexible molecules). NMRPredict is available in Mestrelab's Mnova prediction modules.

15.3.6   Ab Initio Approaches

Another way to calculate chemical shifts is to use Density Functional Theory (DFT). This quantum mechanical approach has been shown to predict chemical shifts well in certain cases. The disadvantage with this and semi-empirical approaches is that they rely on modelling the range of low energy conformers of the structure of interest. Not only is this time-consuming, it is also difficult to achieve in conformationally flexible molecules. Due to its slow performance, it is not a tool in routine use in solving structural problems although it has shown its value in specific cases, particularly where databases are not available for the structural feature/nucleus of interest.

15.3.7   Neural Networks

A neural network (Figure 15.5), in its simplest form, is designed to mimic how the brain learns by using interconnected neurons whose connective weights can be modified by learning.

In the case of NMR prediction, we feed the atom of interest and its environment into the input of a neural network and tell it what the answer is. It adjusts the weights until it gets the right answer. This is repeated using all of the respective atoms of a large number of structures and the weights are adjusted and re-adjusted. This will continue for many hours (or even days) so that the weights give the best answers overall. Normally a proportion of the example data is kept back to use to test the network with something that it hasn't seen before. This is a hugely

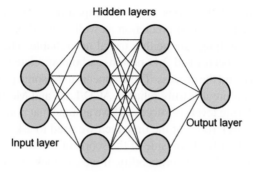

Hidden layers

Output layer

Input layer

Figure 15.5 A neural network.

simplified view of the approach. The design of the network (number of input neurons, number of hidden neurons, number of hidden layers, transfer functions, thresholds for neurons, etc.) is crucial, as is the way that the chemical structure is converted into inputs for the network. Networks take a long time to train but have the advantage of being really fast to use. These tend to be very good at interpolating but poor at extrapolating.

15.5.8 Hybrid Approaches

It may be possible to know if there is greater uncertainty for one approach vs. another. For example, if there are no close HOSE code matches, we cannot trust the result of a HOSE code calculation. Hybrid approaches can perform multiple predictions using different methods and then weight them according to their relative reliability. This can be done for individual protons in a molecule and often leads to more reliable predictions.

15.5.9 Simulation

Spectral simulation is normally provided with proton prediction packages. This takes the predicted chemical shifts and coupling constants and uses them to simulate the appearance of the spectrum. This can be a little misleading as it gives rise to an authentic looking spectrum which may differ considerably in appearance from the experimental one. This is because even small errors in chemical shift or coupling constant prediction may give rise to significant differences in appearance of the signals. Simulation can be useful to try to mimic an observed signal to help calculate the coupling constants when they are not obvious by inspection. Simulation-only software is normally available as part of the NMR acquisition software and may be used to help understand complex splitting patterns observed in real spectra.

15.6 Structural Verification Software

Being able to verify a proposed chemical structure from its NMR spectrum automatically has been a goal for many years. This is particularly true recently since chemists have been making arrays of compounds (tens to thousands of compounds). It is possible to acquire data automatically on large numbers of compounds but it is still a major task to interpret all of the data. Verification software (sometimes called 'Automated Structure Verification' or 'ASV') performs a prediction and simulation and then tries to fit the experimental data to the calculated data. Obviously this approach requires good prediction as well as good data extraction. As you will have seen in this book, these things are neither trivial nor reliable. The latest approaches use a combination of 1-D proton spectra and 2-D proton–carbon correlated spectra to try to use the strengths of ^{13}C prediction to aid the process. There appears to be some promise with this approach but it still has a way to go before it is truly reliable. ACD/Labs and Mestrelab have products that attempt to do this with differing levels of success. If we assume that none of the existing software can ever get the correct answer 100% of the time then we need to ask why we are using it? This is a complex discussion and probably outside of the scope of this book so we will only skim the topic here. Primarily, it comes down to probability. If we can look at everything, then we should look at everything but if we can't, then we have to achieve the most 'bang for the buck'.

In order to use the software, one needs to accept one of the classifications – is the sample 'right' or 'wrong'? By this, we mean does the spectral data agree with the proposed chemical structure? Most decent chemists doing robust chemistry will get the right material most of the time. What we need from the software is to identify potential incorrect samples and these are the ones that we will look at. Since we can only look at a small subset of data, we need to minimise the number of 'false negatives' (the software thinks that the sample is wrong when it isn't) even if this is at the expense of some more 'false positives'. If we imagine 10,000 samples being put through this process and that our chemists make the right stuff 95% of the time, we are trying to find those 5% bad compounds with our limited resources. If we tweak our software to minimise false negatives (say, 5%) at the cost of having more false positives (say, 20%), we can picture the results graphically (Figure 15.6):

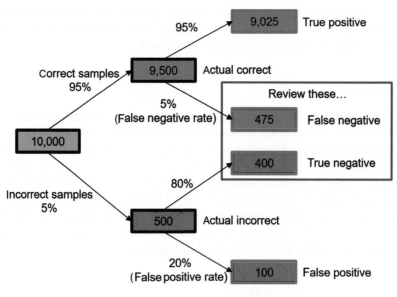

Figure 15.6   ASV outcomes.

By looking at 875 spectra, we have eliminated 400 of the 500 'bad compounds' instead of having to look at all 10,000 spectra. Our collection quality has increased from 95% to 99%. This is one of the most common situations although there may be others where the prior assumption about the proportion of 'good' compounds is lower, in which case the equation changes. This is known as 'Bayesian inference' and is a very important technique in statistical analysis.

15.7   Structural Elucidation Software

This is often called 'Computer Assisted Structure Elucidation' or 'CASE'. Unlike structural verification software, this software is designed to propose structures that may fit the analytical data. The first requirement is for a molecular formula for the mystery compound. The molecular formula is used to set up a combinatorial chemical structure generator which will generate all possible chemical structures for that molecular formula. For all but the simplest structure, this is a massive number. Even $C_6H_8O_6$ gives rise to over 2.5 million structures! It can get even worse if you have sulfur or phosphorus in your MF as these can be incorporated in your molecule with different oxidation states, greatly increasing the combinatorial output. Fortunately, you can narrow down the possible structures a little by excluding certain structural features. You might, for example, want to exclude rings bigger than seven atoms or smaller than four (if this makes sense). You can also use the NMR data to reject certain structural fragments or, even better, use known fragments such as an ethyl group. These fragments are put into 'good lists' and 'bad lists' and this constrains structure generation to a more manageable level.

Going beyond this, it is possible to use the NMR data to constrain or filter chemical structures further. We use as much data as possible to do this and that normally includes: COSY, ^{13}C, HSQC and HMBC in addition to the 1-D proton spectrum. This information gives the software connectivity information that can be used to reject putative structures. It then orders the remaining structures by similarity between the experimental and calculated spectral data. You can browse through the proposed chemical structures to see if you agree with the proposals. What often happens is that you run through the first time and get a bunch of structures that you know cannot be there and so add more structural features to your 'bad list' and iterate through until you find some potentially good structures. The next step is to devise experiments to differentiate between putative structures. This approach has been very useful in the area of natural product structural elucidation.

Another area where this approach has potential is in spotting other possible structures that may fit the data. Even the best spectroscopists sometime become fixated with a structure that fits the data. This software can suggest other possible structures that are worthy of consideration.

A word of warning though. Because the software is using the NMR data to reject structures, it is important that the information is correctly extracted from the data. You will need to help identify overlapping carbon atoms in the ^{13}C data (e.g. para-substituted aromatic systems have two overlapping carbon atoms – you need to let the software know this). Also, HMBC data can be variable in the number of bonds that correlations are observed over. If in doubt, don't pick the peak, otherwise the software may reject possible structures. Lastly, don't forget that we don't 'see' anything except carbon and hydrogen with our standard set of experiments.

In practice, we don't tend to use this approach except when we are faced with a total unknown. There is normally enough information (NMR and chemical) to allow us to propose a small number of chemical structures that are likely and we perform experiments to confirm or refute the possible structures. That said, there are times when we perhaps should use this approach to shake our touching belief that we have the correct structure when there may be another, unconsidered, molecule that would fit the data. We should always accept that we may be wrong!

15.8 Summary

There are many good tools out there to help with processing and assignment of spectra. Choosing the right software for your needs can be a difficult choice as each package brings with it a number of pros and cons. Since vendors offer trial licences, it is worth getting these first before investing in the software. Getting the right software can make your job much easier and quicker. Be warned, though, don't believe everything that the software tells you. It may not have a good model for your molecule and is just 'having a guess'.

16

Problems

In this chapter, we have put together a dozen problems incorporating some of the phenomena we have covered in the text. The first nine are of a hypothetical type where you will be invited to come up with ideas of how you might approach a problem and challenged to suggest realistic courses of action while the final three questions are based on some actual spectra. If you need a bit of inspiration, you might find some in the 'Hints' section. *Enjoy!*

16.1 Questions

Q1. Two possible reaction products from a methylation reaction are shown below. How would you use NMR to positively identify the correct product?

Q2. A purchased compound thought to have the following structure gives a proton spectrum which looks far more complex than would be expected. An HPLC investigation shows it to be a mixture of two components, both of which have the same mass. These are isolated and give rise to two distinct but fairly similar proton spectra, either of which might be considered satisfactory for the proposed structure. What possible explanations might there be for this and how would you investigate the problem to resolve the issue?

Essential Practical NMR for Organic Chemistry, Second Edition. S.A. Richards and J.C. Hollerton.
© 2023 John Wiley & Sons Ltd. Published 2023 by John Wiley & Sons Ltd.

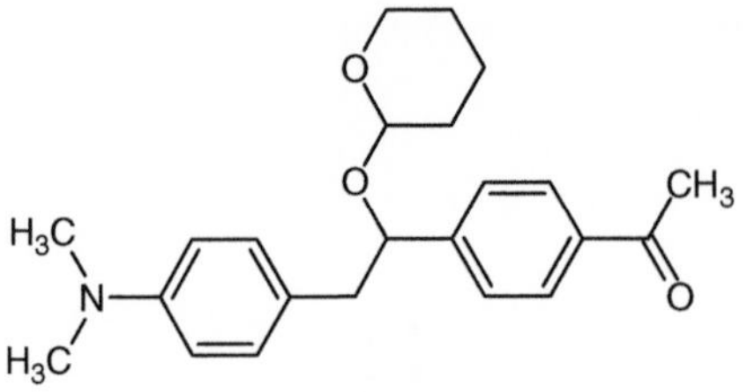

Q3.   The benzylation of the simple triazole shown below could yield one of three distinct products (due to tautomerism of the triazole). Which NMR techniques would you use to positively identify each of them?

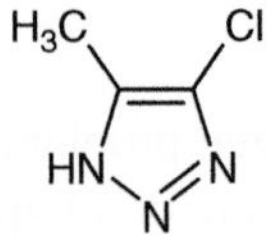

Q4.   It's another benzylation issue… You have benzylated the compound shown below and have separated out three fairly close-running fractions by HPLC which all show the correct mass for a mono-benzylated product. Time is short as you need to get the next reaction on overnight and you have only collected a 1-D proton spectrum on each fraction. What key features in the three spectra might you make use of to assign which fraction is which?

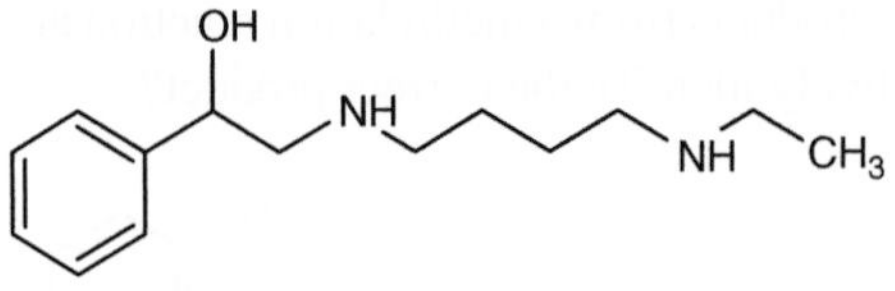

Q5.   You have a sample of the compound shown below but are unsure of its stereochemistry. How can you be certain that it's the 'cis' or the 'trans' isomer?

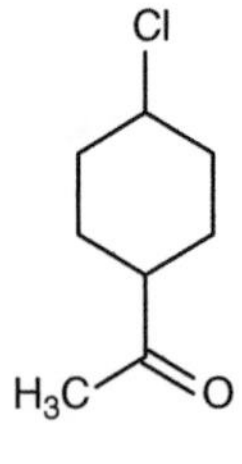

Q6.   You have a sample of a compound shown below but are unsure of the stereochemistry of the chlorine relative to the heterocyclic ring. How would you go about confirming this and what observations might prove pivotal in arriving at a definitive solution?

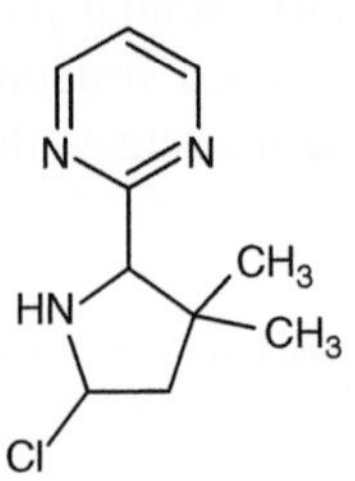

Q7. You have a pair of products from an alkylation which you believe to have the structures shown below.

How would you differentiate them, and could you be certain that your methods would alert you to any other possible reaction?

Q8. A cyclisation reaction has given two products which are shown below. They are similar but have been separated successfully. How would you establish the identity of each?

Q9. The acylation of the compound shown below with acetyl chloride could yield three different mono-acetylated compounds. Consider the key features of each which would enable you to identify each of them.

Q10. You have been attempting the alkylation reaction shown below but the paper you are following is old and lacking in hard evidence regarding the site of reaction. You have isolated one major product that has the correct mass and you have acquired all the NMR spectra that you might need: Spectra 16.1 (^{1}H), 16.2 (^{1}H expansion), 16.3 (^{13}C), 16.4 (HSQC) and 16.5 (HMBC). Use these spectra to define the structure of the product.

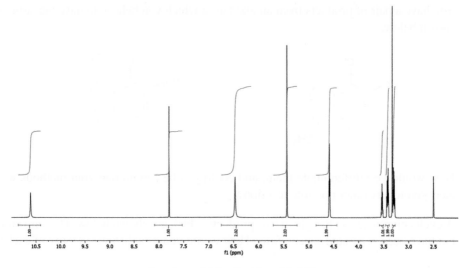

Spectrum 16.1 Problem 10 (^{1}H).

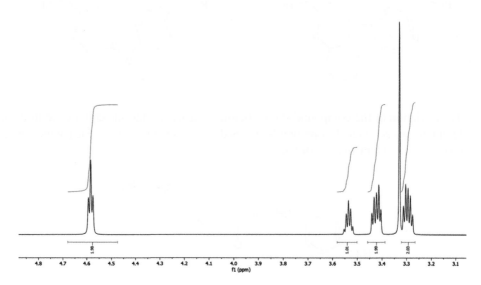

Spectrum 16.2 (^{1}H) Expansion.

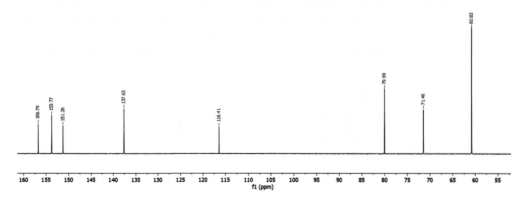

Spectrum 16.3 (^{13}C).

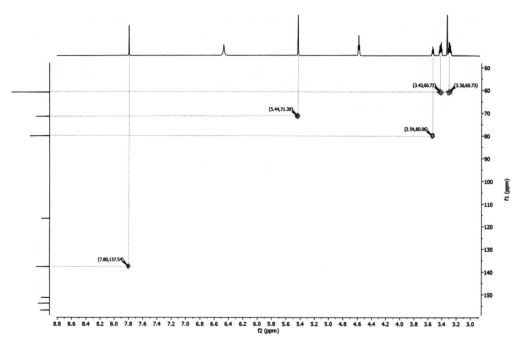

Spectrum 16.4   (HSQC).

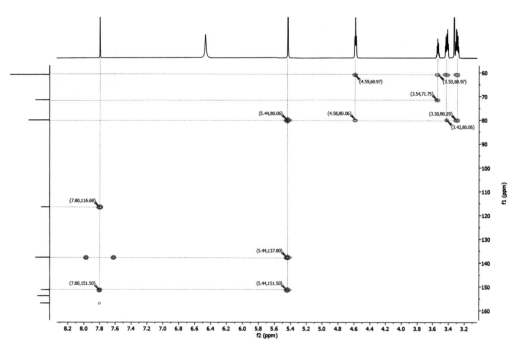

Spectrum 16.5   (HMBC).

Q11.   You have ordered a sample of a compound from a small commercial supplier with the structure shown below. You run a quick mass spec and find a strong molecular ion which is correct for the proposed structure but decide that as you will be using it as the starting material for a long and quite complex synthesis, you'd better check it out thoroughly before proceeding and collect the spectra shown (Spectra 16.6–16.16). Was your cautious approach vindicated in this case?

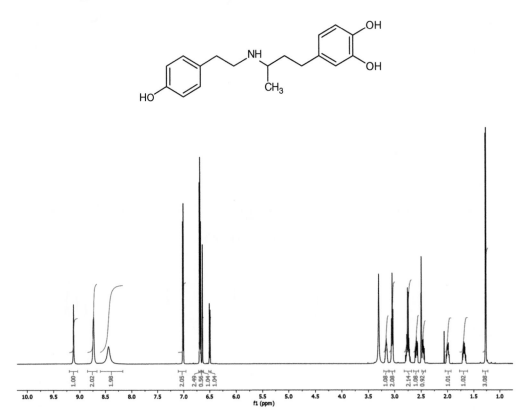

Spectrum 16.6   (^{1}H).

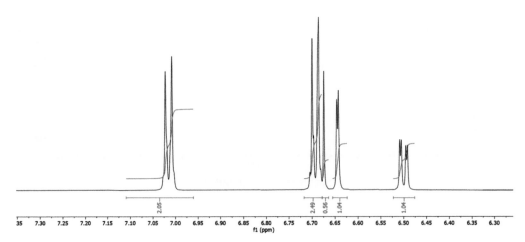

Spectrum 16.7   (^{1}H) Expansion 1.

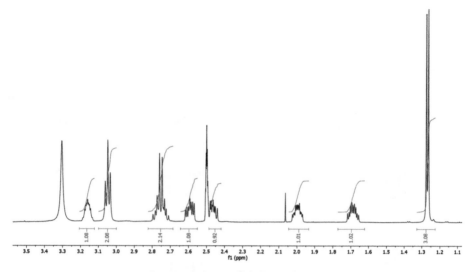

Spectrum 16.8 (^{1}H) Expansion 2.

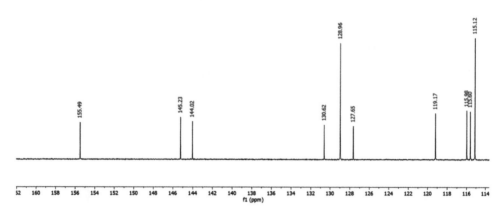

Spectrum 16.9 (^{13}C) Expansion 1.

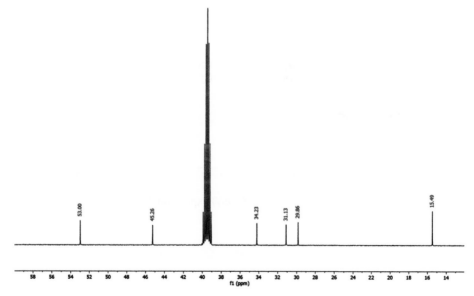

Spectrum 16.10 (^{13}C) Expansion 2.

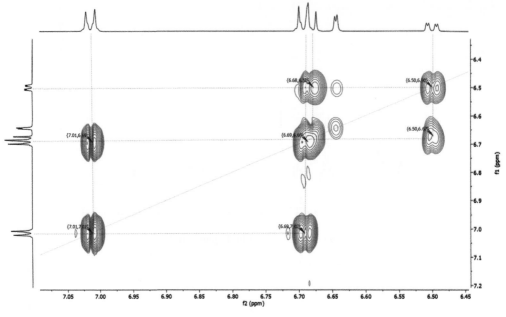

Spectrum 16.11   (COSY) Expansion 1.

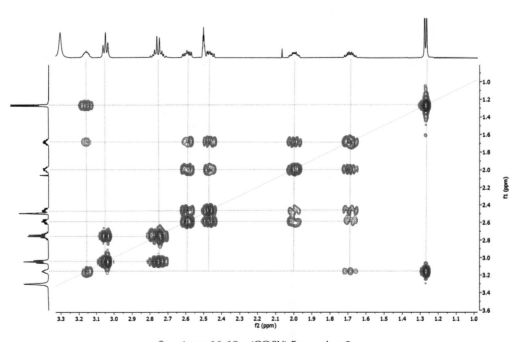

Spectrum 16.12   (COSY) Expansion 2.

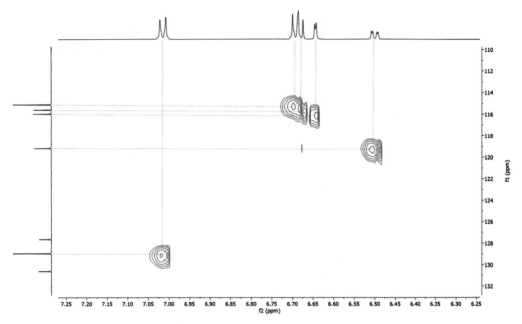

Spectrum 16.13 (HSQC) Expansion 1.

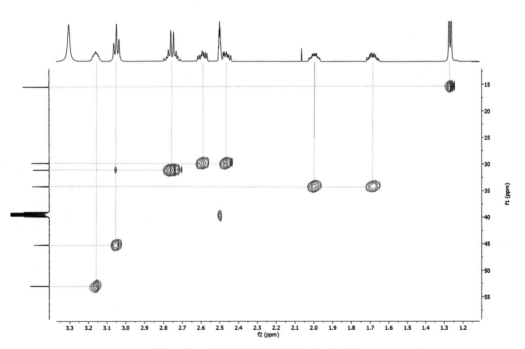

Spectrum 16.14 (HSQC) Expansion 2.

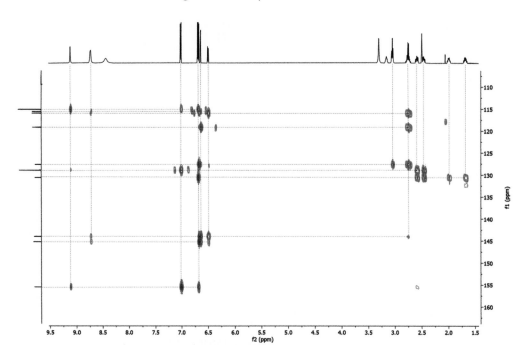

Spectrum 16.15 (HMBC) Expansion 1.

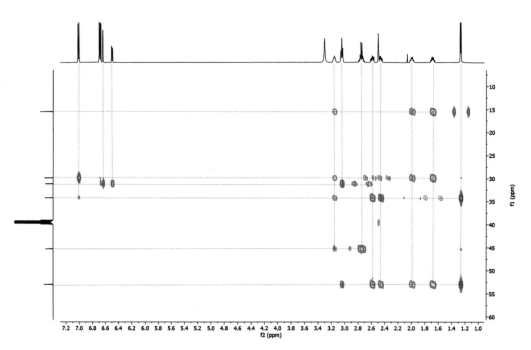

Spectrum 16.16 (HMBC) Expansion 2.

Q12. The final question. (If you've got this far, you deserve a special award for tenacity!) Once again, you have bought in a compound from a small firm you'd never heard of before (Feeling Lucky Pharma?) that were advertising their wares on the internet. The structure proposed is shown below.

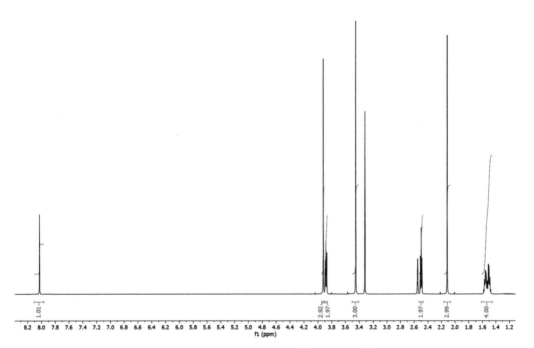

Past experience brings out the sceptic in you (is this really this structure?), so you acquire a full set of spectra despite the mass ion being satisfactory. Spectra 16.17–16.23 may help.

Spectrum 16.17 (^{1}H).

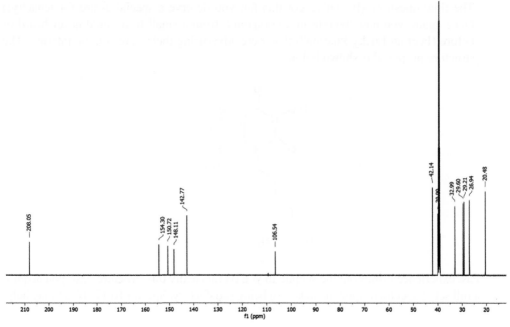

Spectrum 16.18 (^{13}C).

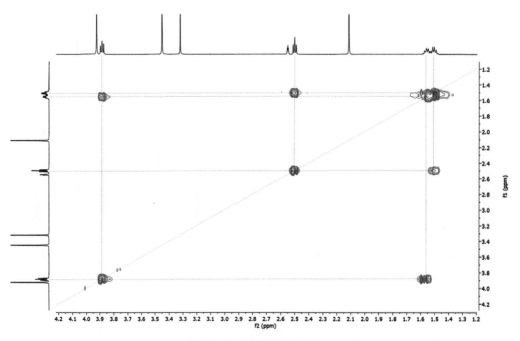

Spectrum 16.19 (COSY).

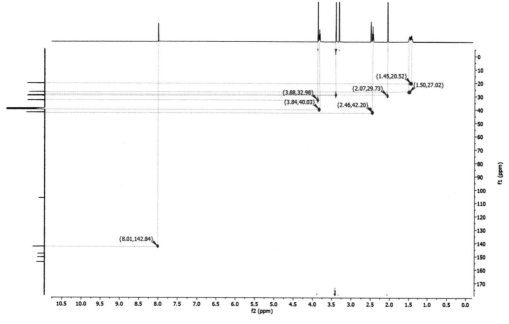

Spectrum 16.20 (HSQC).

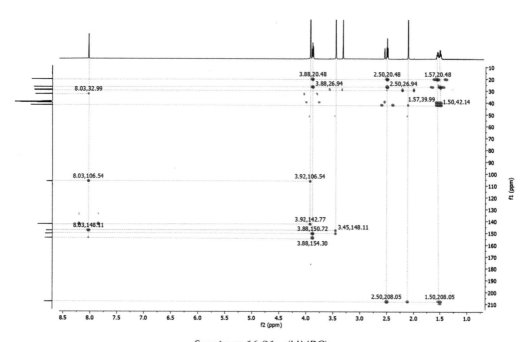

Spectrum 16.21 (HMBC).

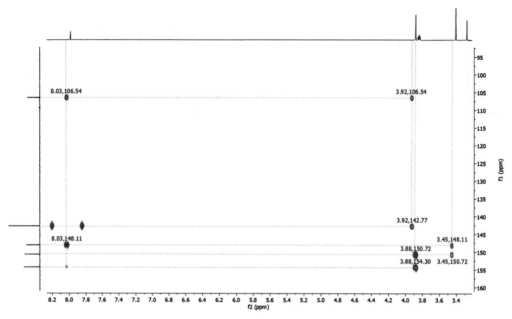

Spectrum 16.22   (HMBC) Expansion.

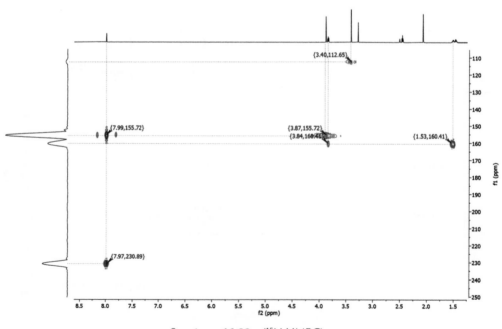

Spectrum 16.23   (^{15}N HMBC).

16.2 Hints

Q1. There are a number of perfectly viable approaches to this problem. Some may be a little quicker and more convenient than others, but the proton chemical shift of the methyl group is *not* diagnostic as it would be quite similar in both isomers.

Q2. Always underpin as much of a structure as you can to avoid the temptations of 'premature speculation'.

Q3. Perhaps not quite as simple as it may seem. Drawing out the three possible regioisomers might be a good place to start.

Q4. Consider the options. Draw them out as in the last question. It might help. Remember that chiral centres can have significant consequences in a spectrum.

Q5. In this case, you might not have to *do* anything other than examine the basic proton spectrum.

Q6. You need to think about the likely conformation of this compound in solution and how it might impact on its proton NMR spectrum – and on how you might use this to your advantage.

Q7. Think about the key corresponding correlations in the two regioisomers. How would they compare?

Q8. When starting to un-pick a set of problem spectra, it's always good to find a sound starting point – some signal which has clear and definitive features (chemical shift, couplings etc.) which you can recognise and use to build an assignment and, ultimately, a solution around.

Q9. It might be helpful to draw out the three likely reaction products. Is rotation about bonds that we draw as single bonds always 'free' on the NMR timescale?

Q10. Here are the predicted ^{13}C shifts of the parent heterocycle. Could be useful…

C2	156 ppm
C3	117
C5	136
C7	151
C9	154

Q11. A first glance at the 1-D proton spectrum is very encouraging. All the right signals are there in the expected places but are they all connected up correctly? Start with a signal that is clearly unambiguous and work through all the connectivities from this signal. A totally rigorous approach is required.

Q12. A positive identification of the N-methyl groups would be helpful. Then check the alkyl chain and see where it fits in.

16.3   Answers

A1.   We hope that you avoided the temptation to draw a conclusion based on the proton chemical shift of the methyl group (note that while the proton chemical shift of an N-methyl group is quite distinct from an O-methyl in the case of alkyl amines, this is not the case in compounds where the lone pair of electrons on the nitrogen is drawn away from the nitrogen – for example where the nitrogen is part of an aromatic system). This is not the case when considering carbon shifts, however. In these potential products, the O-Me carbon would be expected at around 55 ppm whereas the N-Me carbon would be found at around 32 ppm. This difference is certainly highly diagnostic and given the simplicity of the compound, a straightforward 1-D carbon spectrum would be all that is required to do the job. An HSQC spectrum would be an excellent alternative and an HMBC would provide further confirmation in the form of additional correlations to neighbouring carbon(s).

   An alternative approach would be to carry out an NOE or ROESY experiment, irradiating the methyl and to deduce its position from the enhancements observed.

In the case of O-methylation, enhancements of H7 and 9 would be expected, whereas N-methylation would be expected to yield enhancements of H2 and 6. (These would of course be identified by their multiplicities).

   The simplest solution of all, however, probably lies in the selection of the most appropriate solvent (DMSO). In the case of O-methylation, the NH proton would be expected to show a small but distinct coupling to the H2 proton and a slightly smaller (4-bond) coupling to H3. This would give both these protons the appearance of imperfect triplets with H3 being the most upfield of all the aromatic signals. The absence of such extra coupling would be indicative of N-methylation, of course. This method of identification would be highly reliable if two products were to be isolated with one showing this additional coupling and the other, not.

A2.   The first and perhaps most likely potential explanation for the information we have is that the initial sample was in fact a mixture of diastereoisomers since there are clearly two chiral centres in the molecule at C2 and C8. (Note that the introduction of a tetrahydropyran [THP] function, a popular 'blocking group', always introduces a chiral centre into a compound. As is often the case, where there is already a chiral centre present, diastereoisomers will be formed.)

This explanation is highly plausible, as it would indeed give rise to two NMR-distinct species which would be separable on a suitable HPLC system. Nonetheless, as this was a purchased compound and we know nothing of its history, we need to investigate further and a full assignment would be highly advisable.

Starting with the aromatic protons, both our spectra would show two distinct AA'BB' systems for the two aromatic rings. Either the ketone or the amine functions would provide an excellent place to start as these substituents would render H18/20 the lowest field of the aryl signals and H12/14 the highest field of the aryl signals. Having successfully identified these protons, the identification of their coupling partners, H17/21 and H11/15 would be straightforward, either by a simple homonuclear decoupling experiment or by a COSY experiment.

Having unambiguously assigned these protons, we would then be in a position to confirm the regiochemistry of the ABX system for alkyl protons 8 and 9 in relation to the aryl rings. This would be important as the alternative structure below could provide an equally plausible explanation for the two isolated species (i.e. single diastereoisomers of each of the two regioisomers). Confirmation of regiochemistry would be highly recommended in a case like this, in any case.

As far as the methodology of regioisomer confirmation is concerned, NOE experiments targeting the A, the B and the X parts of the ABX systems with the expectations of seeing enhancements of the neighbouring aryl signals would offer a potential line of investigation. A far more convincing approach, however, would be to acquire HSQC and HMBC spectra on both samples where the carbon shifts of the aryl CHs, as well as those of the methine and methylene carbons would be revealed (HSQC) along with key correlations between the alkyl and aryl systems (HMBC).

A3.   The three potential regioisomers are shown below.

Isomer 1   Isomer 2   Isomer 3

Isomer 2 stands out as a great initial target as it has features that immediately distinguish it from Isomers 1 and 3. For a start, the benzylic methylene protons (H6) will only show HMBC correlations to C7 and C8/12, whereas in the case of either Isomer 1 or 3, an additional HMBC 3-bond correlation will be observed to one of the triazole quaternary carbons (C5).

In addition to this, the ^{15}N HMBC experiment should yield chemical shifts for all three of the nitrogens of the triazole ring. N1 would be expected to show a 2-bond correlation from the H6 protons and have a shift of approx. 250 ppm. Though N2 and N5 would give similar shifts (in the region of 330 ppm), it would be likely that these would be sufficiently resolved to enable all three nitrogens to be observed in this isomer. (Check out the useful nitrogen data in Chapter 10)

Discrimination between Isomers 1 and 3 may not be quite so straightforward. The ^{15}N HMBC experiment will only reveal two of the three nitrogens in both cases (N1 at around 240 ppm and N2 at approx. 360 ppm), since N3 will be 4 bonds away from the H6 and therefore out of correlation range in the normal experiment. The problem is that, in both isomers, the shifts of the N1 nitrogen will not be sufficiently different to give an indication of which isomer is which.

Furthermore, the ^{13}C HMBC may also be found wanting – it will not be possible to discriminate between the tetrazole quaternary carbons (4 and 5) in either of these isomers since the methyl protons will show correlations to both of them in both isomers. (Note that while 3-bond correlations are often stronger than 2-bond correlations, relying on the intensity of correlation as an assignment tool in unfamiliar compounds is ill-advised.)

In this case, the NOE or ROESY experiment would be useful since in Isomer 3, an enhancement between the methyl protons (H14) and methylene protons (6) would certainly be expected whereas no such correlation would be observed in Isomer 1. So, to sum up…

Isomer 1: Two nitrogens observed. HMBC correlation between H6 and C5 (shift not definitive). No NOE between methyl and any benzyl protons.

Isomer 2: Three nitrogens observed. No HMBC correlations between H6 and either of the tetrazole carbons. No NOE between methyl and any benzyl protons.

Isomer 3: Two nitrogens observed. HMBC correlation between H6 and C5 (shift not definitive). NOE expected between methyl and benzylic methylene protons as well as a common correlation from H6 and H14 to N1.

A4.   The compound in question clearly has three potential sites where benzylation could take place – at O8, N10 and N15.

Benzylation at O8 would be easily diagnosed by the chemical shift of the benzyl methylene protons but benzylation at either of the nitrogens might be a little more problematic. For a start, the chemical shifts of the benzyl methylene protons would be similar in both N-benzylated regioisomers. Any attempt at deducing the structures of the products based on spin coupling from the remaining NH to neighbouring protons would have a very slim chance of success as amine NHs seldom show spin coupling to alkyl protons (unlike amide NHs which almost invariably do show such couplings).

The key to this problem lies in the chiral centre at C7. Just as this chiral centre would be responsible for the non-equivalence of the two H9 protons, yielding an ABX system for the three protons H7, H9 and H9′, it would also be expected to render the methylene protons indicated (*) non-equivalent as they are very close to the chiral centre. They would therefore be expected to present as an AB system (i.e. a pair of roofed doublets with a geminal coupling of approx. 12 Hz) with a chemical shift typical for an Ar–CH₂–NR₂ methylene. Proximity to the chiral centre is the key.

In the case of the other (N15) N-benzylated regioisomer (shown below), the benzyl group is removed from the chiral centre by a significant distance and as the degree of non-equivalence between geminal protons brought about by a chiral centre diminishes with distance from the chiral centre, it is quite likely that in this regioisomer the methylene protons (*) would approximate to a singlet with a chemical shift quite similar to those in the N10 benzylated compound.

Although by no means a rigorous approach, such observations would be acceptable for differentiating such N-benzylated compounds in the short-term, at least.

[Interesting footnote: in this example, we have picked a relatively simple compound to illustrate a point: that non-equivalence between geminal pairs of protons diminishes with increasing distance from a chiral centre. It is worth bearing in mind, however, that it is possible to encounter compounds that have a preferred conformation in solution such that a geminal pair might be closer in space to a chiral centre than the number of bonds separating them might suggest. It is the 'through-space' relationship that is important here, rather than the through-bond consideration and this phenomenon should be anticipated in compounds having the potential to form intra-molecular H-bonds, and therefore the ability to 'wrap themselves up' nose-to-tail, given a suitable solvent.]

A5. The key to this problem lies in the positive identification of the methine protons (H1 and 4) and on rationalising the couplings that they will exhibit. The 'trans' isomer will perhaps be the more obvious of the pair to recognise.

This isomer will naturally adopt a stable, low energy conformation with both the large groups (–Cl and –COCH$_3$) occupying the less sterically hindered equatorial positions which will lock H1 and H4 into permanent axial environments. This rigid stable 'chair' structure will yield predictable couplings between all H2/6 protons and H1 and between all H3/5 protons and H4. Both H1 and H4 will be of similar appearance with chemical shifts of approx. 4 and 2.5 ppm respectively. They will both show two large axial–axial couplings of approx. 12 Hz and two smaller equatorial–axial couplings of approx. 3–4 Hz. Thus, in appearance, both will be a triplet of triplets.

In the case of the 'cis' isomer (shown below), the appearance of the methines will not be so clear cut. There will be something of a steric conflict between the two large groups over which will get to occupy the lower energy equatorial position, for most of the time at least.

The natural expectation would be for one of the groups to 'win' this 'steric sumo contest', the prize being permanent occupancy of the less hindered, lower energy equatorial position. The consequence of this would be that the other substituent group would be forced into the more energetic axial position. If this were to be the case, we would expect to see one clear axial methine proton with multiplicities as described above and one clear equatorial methine proton where all couplings from the neighbouring methylene protons would be approximately 4 Hz (all couplings would be either equatorial–equatorial or axial–equatorial. Check the Karplus curve).

However, in cases like this, the outcome is unlikely to be decisive as the lowest energy conformation of the cyclohexyl ring will not be that of the classic 'chair'. While we would encourage the use of molecular modelling kits, it must be recognised that these present a very static picture of a molecule. This is fine in cases where one clearly defined conformation predominates but it must never be forgotten that the true picture is a dynamic one and in cases like this one, the couplings we observe will be the resultant time-averaged sum of the predominant conformations with interconversion between them being fast on the NMR timescale.

So, to sum up, in the case of a 'trans' isomer, expect methine protons to be clearly axial with two large and two small couplings while in the case of a 'cis' isomer, expect at least one, and possibly both, methines to present with smaller 'hybrid' couplings, the clear distinction between axial–axial and axial–equatorial couplings diminished.

(Reminder: it is worth remembering that as we saw in Section 6.7, in the majority of cases, a proton in an equatorial environment will have a more downfield chemical shift than it would if it were to be in an axial environment. This difference is typically 0.5–1.0 ppm and can be usefully diagnostic. This is a general observation and it should be noted that in the case of certain substituents, the trend can be reversed as specific anisotropies may be over-arching.)

A6. Sometimes, we can use a specific feature of a molecule to help us extract the information needed to arrive at a sound solution. In a case like this, molecular anisotropy would prove to be a very useful tool indeed.

As a 'thought experiment', try to envisage what the spectrum of the compound might look like – without the pyrimidine heterocycle in place. Though this compound would still have a chiral centre at C2, making the methyl groups, 7 and 8 non-equivalent and therefore likely to have different chemical shifts, this difference would be likely to be small. The addition of the heterocycle would change things dramatically.

In order to minimise steric interactions, the molecule would have to take up a conformation to minimise energy where the pyrimidine ring would be out of the plane of the paper (i.e. vertical). This would mean that the methyl group on the same face of the five-membered pyrrolidine ring as the pyrimidine (methyl 7 as drawn above) would be held in the shielding zone of the pyrimidine and would be likely to have a chemical shift significantly upfield of its partner, methyl 8. The expected shift would be about 0.6 ppm for H7 protons and about 1.5 ppm for H8. Note that there is an attractive logic here. An aromatic or heterocyclic ring can exhibit either anisotropic shielding or deshielding, depending on the region of the ring that is forced into close contact with the group in question. The fact that we would observe an abnormally upfield methyl in this molecule indicates that the phenomenon is a shielding one. This implies that one of the methyls must be held over the face of the pyrimidine ring (i.e. on same face of pyrrolidine) and all further deductions flow seamlessly from this.

Having understood the significance of this anisotropic shift, the way ahead would be clear enough – NOE experiments, targeting both methyl groups. One of these would be likely to show a clear enhancement across the ring to H2a and this would make deduction of the chlorine stereochemistry relative to the pyrimidine, quite straightforward. (Remember a positive NOE enhancement is what we would want to see. The lack of an NOE enhancement is not good evidence to support a structure.)

(Interesting footnote: Were the molecule to have a benzene ring in place of the pyrimidine, then a direct NOE from the ortho protons of the benzene ring would be well worth trying and would probably be successful but in our example, N10/14 prevent this and H11/13 would be too remote to yield a result so making use of the methyl groups as outlined would be necessary.)

A7.   In cases like this, you need to consider the applicability of the various NMR experiments at your disposal in order to assess which will give you the clear distinction between the potential products that you need.

It soon becomes clear that the NOE experiment has nothing to offer in this case. In both compounds, targeting methyl (10) can only yield enhancements of H2.

The ^{15}N HMBC experiment might then suggest itself and initially appear attractive as the chemical shifts of N1 and N3 would certainly be distinct – but there is a problem. The bonding to the nitrogen changes on methylation so that in both cases, the N1 nitrogens will have similar values (approx. 150 ppm) and so will the N3 nitrogens (approx. 250 ppm). There would be slight differences – but they would be far too slight to predict with any confidence. N1 and N3 will also show the same correlations (from H2 and H10) in both cases and there will be no unique corroborating correlation in either compound to allow the distinction to be made. So the ^{15}N HMBC would not be helpful in this case.

The 'silver bullet' in this example would be the ^{13}C HMBC experiment. In both compounds, the methyl H10 protons would show two, 3-bond correlations. The chemical shifts of the C2 carbons would be similar in both compounds (expect in the region of 140–145 ppm) but the quaternary carbons C4/C5 in the first regioisomer* and C4/9 in the second would show a significant difference. As C4/5 is flanked by two nitrogens, it would show an absorbance somewhere around 150 ppm whereas the corresponding quaternary, C4/9, in the second isomer would have a shift in the region of 125–130 ppm. This difference is significant and reliable and would confer a high degree of confidence on the assignments. *Note different numbering in the two isomers.*

In the event of the reaction giving rise to the unexpected N-methyl quaternary compound shown below, we would be alerted in two important ways.

Firstly, we might expect to observe broadening of signals, particularly in the five-membered ring, in either proton or carbon domains, or maybe in both due to tautomeric interconversion as the NH migrates between N1 and N3. The exact observation would of course depend on the rate of this process on the NMR timescale.

The second 'safety net', would be the common correlation between H11 and H12 and C8 which would of course not be seen in the regioisomers previously discussed.

A8. Distinguishing these regioisomers from each other might prove to be quite challenging as the two structures are so similar.

The one stand-out feature that offers a 'way-in' to this problem, is the difference between a -CH$_2$-Br and a -CH$_2$-Cl in terms of both ^{1}H and ^{13}C chemical shifts. There will be a proton shift difference of about 0.3–0.5 ppm between these protons with the bromo-species being at higher field. An HSQC spectrum on each would give greater confidence with the carbon shifts being separated by around 10 ppm, the bromo-species expected at around 35 ppm and the chloro- species at around 45 ppm.

While these observations may seem of limited interest since both regioisomers have both these groups, in fact, they offer two potential solutions to this problem. Starting with the first compound, the H17 protons would be expected to show a definite correlation to N6 in a ^{15}N HMBC experiment. N6 would also be likely to show a correlation from the H4 protons (which would have been easily identified from a simple 1-D proton spectrum).

The second compound would also show a similar correlation (in this case from H19 to N13), but no further correlation from the H4 protons which would be correlating to the N6 nitrogen. Though the pyridyl nitrogens would give very similar shifts to each other, sufficient difference between them to allow definite resolution would be expected.

A second possible line of attack would be the regular ^{13}C HMBC experiment. In the first compound, the H17 protons would show a strong 3-bond correlation to C8. This would allow positive identification of H8 via the HSQC experiment. H8 could now be used to register a positive identification of C10 which would also show 3-bond correlations from both the H2 and H4 protons. Only the H2 protons would correlate to the corresponding C11 carbon in the second regioisomer.

This example indicates how, sometimes, problems have to be solved by indirect reasoning as there may be no obvious one-shot solution available.

A9. This problem demonstrates how important it is to identify potential pitfalls and factor them into your thinking as you approach a problem. Assign in haste (incorrectly), repent at leisure! A good place to start would be the 1-D proton spectrum of the starting material. You should assign this fully, achieving a high degree of confidence in the assignment of each of the proton species. H2/6 would inevitably be close in chemical shift to H17/19 as would H3/5 and H16/20 but these could be quickly and unambiguously assigned by using an NOE technique targeting the methyl protons (H12). This would give a positive identification of H3/5 which would pave the way to full identification of all the aryl protons. H8 and H14 would be identifiable by proton shift and while H9 and H13 would be close and possibly overlapped to some extent, particularly on a lower field instrument, this would not present any problem in terms of assignment.

Identification of H10 would be straightforward as this signal would be out on its own at just below 4 ppm.

Starting with the –O acetylated product, a downfield shift of about 0.5 ppm would be expected for H10 relative to the starting material. This would require little more than careful use of the relevant proton chemical shift prediction table. It is important to identify this chemical shift change as the most important feature in the case of this O-acetylated species – and here's why... When a reaction with an acid chloride is carried out, a mole of HCl is released for every mole of product produced. As our compound contains basic nitrogens, we need to be aware of the possibility that these may become protonated. So if you were to seize on a downfield shift of H2/6 or of H17/19 as indication of reaction at N7 or N21, while failing to note the shift change in H10, you could easily find yourself backing the wrong horse! A safe way to side-step this problem would be to add a drop of D$_2$O solution saturated with sodium carbonate to neutralise the liberated HCl prior to running the spectra of your products.

N-acetylation at N21 would be easy to spot with the H17/19 protons moving downfield by as much as 1 ppm. No significant shift of H10 or H2/6 would be noted (basification assumed). Note that the -NHCOCH$_3$ group deshields ortho positions on a benzene ring both by withdrawal of the nitrogen lone pair by the carbonyl function and also by 'through-space' (anisotropic) deshielding.

Whereas acylation of a primary amine gives rise to a secondary amide, which will always be unlikely to give rise to restricted rotation problems as the 'trans' conformer is invariably highly predominant, acylation of a secondary amine gives rise to a tertiary amide – and that's a different story.

In this compound, restricted rotation about N7–C13 would be a certainty. This would give rise to probable doubling of signals in the region of the amide bond. Doubling of signals for H2/6, H8 and H23 with intensities reflecting the relative contribution of each of the rotamers would be expected. If in any doubt, confirmation of the presence of rotamers could be easily achieved either by running a proton spectrum at high temperature in DMSO (suggest 120°C) which would lead to coalescence of rotameric signals. Alternatively, an SPT experiment where spin is transferred from one rotameric signal (e.g. one of the H23 signals) to the corresponding signal in the other rotamer could be employed. This would certainly be the preferred method in cases where the products are likely to be thermally unstable. Both these topics have been previously discussed in the text.

A10. There would certainly be plenty of possible products from this alkylation! What is needed is a clear expectation of spectral features that we might expect for each of the potential products. Considering each of these possible compounds, it soon becomes clear that the NOE experiments would be of little help in identifying any of them and we will have to look elsewhere for answers. The key to the problem lies in the HSQC and HMBC data.

Firstly, we must identify the proton and carbon signals for the methylene that is attached to the heterocycle. In the proton spectrum, this is clearly the singlet at about 5.4 ppm. Now, by using the HSQC spectrum, we can confirm that the shift of the carbon bearing these protons is 71.4 ppm. This shift immediately rules out the possibility of -O-alkylation, as in this case the methylene in question would be between two oxygen atoms and therefore have a carbon shift in the region of 100 ppm.

Without looking at each case in complete detail, it is immediately clear that alkylation at N1, N8 or N11 will not show any correlation from the methylene protons to C5 whereas alkylation at N4 or N6, will. Recognition of this is a big step to a solution for this problem. The next thing we need to do is to find the chemical shift of C5, so it's back to the HSQC again. H5 is easily identified (sharp singlet at 7.8 ppm) and this proton correlates to a carbon at 137.6 ppm. Now if we look at the HMBC spectrum, we can see at a glance which carbons show correlation to the methylene protons of the alkylating group (singlet at 5.4 ppm as noted earlier). They are: 80 ppm (correlation to the methine carbon in the alkylating agent itself), 137.6 ppm, which we know is the shift of C2 and 151.3 ppm which has to be C9 on chemical shift grounds. (Note the shift of C3 in the starting material, given in the 'Hints' section, was 117 ppm). We can therefore be confident that we have the regioisomer shown below:

A full assignment of the proton and carbon spectra is shown below: (Note: numbering differs from starting material.)

H2, 7.8 ppm (singlet)	C2, 137.6 ppm
	C4, 116.4 (correlates to H2)
	C5, 156.8 (tentative – no correlations)
H6, 11.1 (broad singlet)	
	C7, 153.8 (tentative – no correlations)
	C9, 151.3 (correlates to H2 and 10)
H10, 5.4 (singlet)	C10, 71.4
H12, 3.5 (multiplet)	C12, 80.5
H13/17, 3.3 and 3.4 (two multiplets)	C13/17, 60.9
H14/18, 4.6 (triplet)	
H16, 6.5 (broad singlet)	

It is interesting to note that, in this example, a ^{15}N HMBC would not have been as useful as might be expected because while the chemical shift of N1 and N3 would be very distinct, in the event of alkylation at N3, the chemical shifts of these two nitrogens would trade places, i.e. the alkylated nitrogen is always the less deshielded one.

Bonus marks if you have noticed that the signals for H13/17 are more complex than you might have expected. This is because these protons are diastereotopic and therefore not equivalent. (This phenomenon has been discussed in detail in an earlier section, *'Enantiotopic and Diastereotopic Protons'*).

A11. Starting as usual with the 1-D proton spectrum, we can immediately see many of the features we might reasonably expect, given the proposed structure. The two distinct spin systems for the two aryl rings are very much as they should be with one of them exhibiting the classic AA'BB' system of a 1,4 disubstituted ring with a shielding substituent at one end and a fairly neutral substituent para to it (6.70 and 7.02 ppm). The second aryl ring also exhibits a classic pattern, that of a 1,2,4 tri-substituted ring with one neutral and two shielding groups (a large doublet at 6.70 ppm which underlies one half of the AA'BB' system for the other aryl ring, a fine meta-coupled doublet at 6.65 ppm and a doublet of doublets at 6.50 ppm). All these couplings and shifts are entirely in keeping with the proposed structure.

Below the aryl signals, between 9.2 and 8.3 ppm, we see three exchangeable signals (check out the HSQC spectrum and you will see no correlations from these signals in the carbon domain). They are of varying degrees of broadness and two of them integrate nearer to two protons than to one, indicating a possible association with a water molecule. Nonetheless, assigning them to the three phenolic protons and possibly the amine NH in the proposed structure is entirely reasonable and this can be further probed via the HMBC spectrum since they are sufficiently sharp to show correlations to some of the aryl carbons. We will return to these in the final assignment.

Moving on to the alkyl region of the spectrum (3.5–1.0 ppm in this case), we see immediate evidence of a chiral centre in the molecule with obvious examples of what appear to be non-equivalent, geminally-coupled protons on the same carbon atoms, e.g. the multiplets at 2.60 and 2.45 ppm as well as the pair at 2.00 and 1.70 ppm. This is easily confirmed by looking at the HSQC spectrum which is always an excellent tool for linking non-equivalent protons to the carbon they are bonded to.

In Chapter 5, we advised working through the spectrum from left to right. However, in some cases, it is often more convenient to 'mix-and-match' your approach and this is one of them. Having satisfied ourselves with the aryl region, there is absolutely no reason not to switch our attack to the alkyl region and use a definite signal to open a 'second front' in the battle. The methyl doublet would certainly be a signal to work on via the COSY spectrum and by using this approach, we can easily establish the coupling relationships of the multiplets to each other. In this way, we can see that the methyl doublet at 1.28 ppm couples to the broad multiplet at 3.17 ppm which, in turn, couples to the geminal pair at 2.00 and 1.70 ppm and this pair couples to the second geminal pair at 2.60 and 2.45 ppm and this is the end of the coupled chain so this last pair must be joined to an aryl ring as in the proposed structure. Remember that the intensity of the correlations to a geminal pair is seldom equal in a COSY spectrum as the size of the vicinal couplings is unlikely to be equal. Having come so far with nothing but green lights all the way, it might be tempting to conclude that this structure fits the spectra perfectly but to be completely confident, we must prosecute it further...

So looking at our proposed structure again, we should see 3-bond correlations in the HMBC spectrum from this geminal pair (2.60 and 2.45 ppm) to the carbons associated with the meta-coupled aryl doublet (6.65 ppm) and the doublet of doublets (6.50 ppm) as well as a 2-bond correlation to a lower field aryl quaternary carbon. We find these from the HSQC of course and note that their values are 116.5 and 119.7 ppm respectively. Now is the time to check this from our HMBC...and we find that this is not what we see at all!

We see only a single 3-bond correlation to carbon at 129.6 ppm and a further correlation to a quaternary carbon at 131.2 ppm. This is clearly wrong and we must reconsider the structure. Looking for the 129.6 ppm in the HSQC spectrum, we see that it is associated with one half of the AA'BB' system in the *other* aryl ring! How can this be? Time for a spot of creative thinking... What if the two aryl rings were swapped around? Would this structure fit the bill?

It does! A full proton and carbon assignment is given here

	C1, 145.8 (correlates to H3 and 6)
	C2, 144.6 (correlates to H3, 4 and 6)
H3, 6.69 ppm	C3, 116.2 (correlates to H8 and 4)
H4, 6.51	C4, 119.7 (correlates to H9)
	C5, 128.1 (correlates to H9 and 10)
H6, 6.65	C6, 116.5 (correlates to H9, 7 and 4)
H7, 6.75	
H8, 6.75	
H9, 2.80	C9, 31.7
H10, 3.05	C10, 45.9
H12, 3.17	C12, 53.6
H13, 2.00, 1.70	C13, 34.8
H14, 2.60, 2.45	C14, 30.4
	C15, 131.2 (correlates to H13, 14 and 17/19)
H16/20, 7.02	C16/20, 129.6
H17/19, 6.70	C17/19, 115.8
	C18, 156.0 (correlates to H16/20, 17/19 and 22)
H21, 1.28	C21, 16.1
H22, 9.12	

A12. In this example, after a fairly quick look at the 1-D proton spectrum, we can again opt for the 'mix-and-match' approach and move on to the HSQC spectrum which we can use to identify the methyl groups and the alkyl chain. Confirming the C-methyl of the ketone is straightforward as its proton shift is bound to be similar to that of acetone and can be seen at 2.06 ppm and 29.3 ppm (^{13}C) from the HSQC. Still looking at the HSQC, the other two methyl singlets can be seen at 3.40 ppm (^{1}H) and 28.9 ppm (^{13}C) and 3.88/32.7 ppm. These can both be confirmed as N-methyls from the ^{15}N HMBC showing nitrogen correlations to these methyl protons of 113 and 156 ppm respectively.

The proton singlet at 7.98 ppm is a good place to begin making the connections that will eventually tie the structure together. Its corresponding carbon shift of 142.5 ppm is exactly as expected for a CH between two nitrogens in a heterocycle of this type.

Turning to the HMBC, we can see that this CH shows strong correlations to quaternary carbons at 106.3 and 147.8 ppm. We can also note a common correlation to the 106.3 ppm carbon from the methyl protons at 3.88 ppm. This is clearly a problem for the proposed structure and indicates that this methyl group must be attached to the other nitrogen in the five-membered ring.

Furthermore, we can see a common correlation from the other methyl protons (3.40 ppm) and the proton singlet at 7.98 ppm with a quaternary carbon at 147.8 ppm. Taking note of all these observations, and after a bit of creative thinking to re-arrange methyl groups, suggests the structure below as a good candidate for the compound. We still need to confirm the location of the alkyl chain and confirm that there is a correlation between H10 and C2 and C9. A full assignment, including ^{15}N correlations, is given below.

		N1, 160 ppm (correlates to H11)
	C2,150.4 ppm (correlates to H10 and 17)	
		N3, 113 (correlates to H17)
	C4, 147.8 (correlates to H6 and 17)	
		N5, 231 (correlates to H6)
H6, 7.98 ppm	C6, 142.5	
		N7, 156 (correlates to H6 and 18)
	C8, 106.3 (correlates to H6 and 18)	
	C9, 154.0 (correlates to H10)	
H10, 3.82	C10, 39.8	
H11, 1.51	C11, 26.7	
H12, 1.45	C12, 20.2	
H13 2.45	C13, 41.9	
	C14, 207.8 (correlates to H12, 13 and 20)	
H17, 3.40	C17, 28.9	
H18, 3.88	C18, 32.7	
H20, 2.06	C20, 29.3	

16.4 A Closing Footnote

You might be thinking that some of these questions, particularly the last three, are a little contrived and of course, you'd be correct. But having said that, they are nonetheless entirely typical of problems that we see on a regular basis. You might be wondering why you would feel the need to confirm a structure, when the supplier has given a clear indication of this? After all, when you buy a bottle of shower gel at your local supermarket, it wouldn't cross your mind for a moment that it might actually be drain cleaner…

Unfortunately, this is not the case when it comes to the purchase of small quantities of organic compounds for research purposes. Some of the small companies in this market buy up collections of compounds from university research projects and sell them on to interested parties. These small companies take structures as read and lack the resources to check their compounds out properly. They are 'sold as seen' and it is very much a case of 'Buyer beware!' You might be lucky, so take a punt by all means but we would advise you to check out the merchandise carefully before committing time and effort to any subsequent chemistry effort.

17

Raising Your Game

We hope you've found this book useful and informative so far – and maybe even somewhat enjoyable (?)! We've looked at a great many aspects of NMR spectroscopy but have always tried to maintain a definite focus on the techniques as practical tools for problem solving in either an industrial or an academic environment.

But if the title of this chapter seems a little strange, we hope its relevance will soon become clear. (We were thinking of calling it something like 'Zen and the Art of NMR Interpretation', but considered it a little pretentious – and that would never do! But seriously...) NMR problem solving has certain similarities with playing chess. For example, just as it doesn't take very long to learn the rules of chess and the movements of the pieces, so gaining the basic ideas behind the various NMR experiments we have looked at is relatively straightforward. The analogy is quite an apposite one and so it continues – in contrast, it takes a great deal of time and study to become a competent chess player and becoming competent in the field of structural elucidation by NMR is a similarly arduous journey.

So the purpose of this chapter is to attempt to give some insight into the actual mechanism of problem solving; how to navigate a safe passage between the icebergs of ambiguity to the safe harbour of a robust solution to a problem. We'll be concentrating on the thought processes that we run through when confronted by a problem and on how we build information up layer on layer in an iterative manner, drawing data from the spectra that are most relevant to each aspect of a proposed structure.

17.1 Spotting the Pitfalls

By now, you might be wondering how to develop these skills and we regret that we can't offer any quick fixes. It certainly isn't about committing large amounts of chemical shift and coupling data to memory and it isn't even about cultivating instant photographic recall of various spin systems, though both these abilities are undeniably useful. It's more about 'seeing' your way through to a solution. What information do you need in order to propose a solution for that unrecognised reaction product and which technique(s) do you use to obtain it? It's also about identifying potential pitfalls of a proposed problem-solving strategy *before* falling into them.

Essential Practical NMR for Organic Chemistry, Second Edition. S.A. Richards and J.C. Hollerton.
© 2023 John Wiley & Sons Ltd. Published 2023 by John Wiley & Sons Ltd.

Always ask yourself the questions – 'What if…?' and keep them in mind. Let's consider just a small selection of these pitfalls…

17.2   The Wrong Solvent

The very first issue to be avoided would be the selection of an unsuitable solvent. The characteristics of the main NMR solvents have already been discussed at some length earlier in the text but it's worth re-visiting this topic. There seems to be a common misconception in the chemistry community that samples can't be recovered from DMSO solution (because of its low volatility) without a great deal of effort. This isn't true. We've found that 0.5 ml NMR samples in this solvent can be blown down to dryness in a few hours by just playing a stream of nitrogen over them in a suitable vial. While the use of the relatively volatile D4-Methanol as an alternative for a polar compound may seem attractive, the loss of information that you might experience may take longer than this to put right.

As we saw in the first question of Chapter 16, the use of DMSO rather than D4-Methanol could enable the site of alkylation on an indole to be determined directly from the 1-D proton spectrum, neatly side-stepping the need for further work in the shape of either ¹³C or NOE studies. (Remember the diagnostic NH couplings to the H2 and H3 protons that could be so useful and available when the compound is run in DMSO?)

Our advice would be – always consider the information you might be throwing away by running a sample in D4-Methanol before committing your sample to it. Consider, for example, the compound shown below.

How many protons of this compound do you think may be prone to full, or at least significant partial exchange with deuterium if it were to be dissolved in D4-Methanol? Phenolic protons H7/8, amine protons H14 and thiol H16 would certainly be obvious candidates that would exchange pretty rapidly. (Note that observing H16 would be particularly important in this molecule as di-sulfide dimers can form easily. Detection of a thiol proton, which would be likely to be quite sharp and distinct would be excellent reassurance that this has not happened. As a point of interest, H16 in this compound might be expected to present as a 4-line multiplet on account of the chiral centre at C13 which might very well bring about different couplings between H16 and the two non-equivalent H15 protons.)

Less obviously perhaps, would be the loss of H10 and H13, both of which would be prone to exchange via the keto-enol tautomeric route if the sample were to be left standing for any length of time in D4-Methanol. Less obvious still is the possibility that the aryl protons H2, 4 and 6 (particularly H2) might well exchange via the keto-phenol tautomeric route. (We have observed this in aryl compounds with two phenolic groups meta to each other.) So, running this compound in D4-Methanol wouldn't leave you with a lot to look at! This might be a rather extreme example but it does make the point fairly dramatically.

As a footnote to this section, it is worth noting that if you *have* dissolved your compound in D4-Methanol only to find that you are experiencing unwanted deuterium exchange, it is possible to recover the situation and to return your compound to its protonated form. The method is simple enough but is somewhat time-consuming. Firstly, blow the solution down to dryness using a stream of dry nitrogen. Then, re-dissolve the sample in ordinary methanol and shake vigorously for a few minutes. Warming the sample periodically during this process will aid the back-exchange. Now blow the sample down to dryness once again and re-dissolve in the solvent of your choice. Hopefully, the deuteriums will have been removed and re-exchanged for protons and no harm should have been done.

We have stressed the potential dangers associated with the use of D4-Methanol as an NMR solvent, as they are predictable and therefore avoidable. Of course, there may well be problems associated with other NMR solvents when used to dissolve certain classes of compounds and there is often a certain amount of trial-and-error required to identify the solvent that gives the best results (i.e. best separation of signals in crowded regions of the spectrum) in any given situation.

17.3   Choosing the Right Experiment

We have already seen in the section on the NOE experiment how this technique can provide rich pickings in terms of pitfalls with potential bogus enhancements relayed via exchangeable protons, etc. so we won't re-visit them again here but there are plenty of other potential issues you need to be aware of! When considering any potential spectroscopic problem, you should always bear in mind the features of the individual compound rather than just the class of compound. For example, given the two possible structures below, differentiating between them would be easy enough by NOE experiment and this would be the preferred approach (A = some unspecified group).

Assuming sufficient chemical shift separation, irradiation of the ortho protons, H11/15 would be likely to give an enhancement of the H6 proton in the case of the first compound but there would obviously be no defining enhancement in the second compound. Simple. But what would happen in similar examples where C2 is substituted with groups of varying size? The NOE method described works because both the ring systems of these compounds tend to lie in-plane with each other (allowing favoured conjugation between the ring systems) so the inter-atomic distance between H11/15 and H6 enables NOE enhancement between these protons, to occur. But steric interaction between any C2 substituent will be likely to force the benzene ring out of the plane of the indole ring to some extent, leading to a greater inter-atomic distance between H11/15 and H6 and thus, to a weaker NOE. In the case of a large substituent such as -*t*-butyl group at C2, it is quite likely that no NOE would be observed at all.

So, while the use of the NOE technique to differentiate isomers of this type is quite sound, it comes with the proviso that the technique may be unsuitable for analogues that are C2 substituted. Recognising this potential pitfall and side-stepping it by approaching the problem with a different strategy would prevent embarrassment!

As a final example of how an investigation can 'come off the rails', let's consider the benzylated triazole shown below. The predicted product is the one shown but you want confirmation that the benzyl has in fact gone on the central nitrogen (9). (Remember from Question 3 in Chapter 16 that tautomerism of the triazole NH gives rise to the possibility of alkylation of any of the three nitrogens.)

So, what are the potential problems that might be encountered here and how can we avoid, or at least mitigate them as far as possible? Before answering these questions, we must first consider the options open to us. Clearly, there is no NOE possibility here as even if alkylation was to have occurred at N8, there would be no chance of any significant NOE between CH_2 (12) and either of the furan protons (H2 and H3).

Two other possibilities suggest themselves. Firstly, the HMBC experiment. In the case of the product as drawn, the CH_2 (12) protons could only correlate with the quaternary carbon (13) and carbons 14/18, so only two correlations would be seen from the benzyl protons, whereas benzylation at either of the other two nitrogens (8 or 10) would lead to a third correlation to either C7 or C11. Sound strategy? Yes – but what are the dangers? What would happen if C7, or C11 were to be either accidentally equivalent in shift to C13 or very close to it? The danger would be that the third correlation may be missed and therefore, incorrect deductions might be drawn from the data. To mitigate this risk, we could predict the chemical shifts of the possible products (using ACD prediction software, for example) to evaluate the potential for the chemical shifts of these carbons to be close. We could also run the 1-D ^{13}C experiment in addition to the HMBC, as the 1-D experiment allows far greater resolution of carbon signals so that the dangers of failing to note accidental equivalence of key signals would be much reduced. This would alert us to the likelihood of the HMBC experiment failing to differentiate the two quaternary carbons in question.

Another strategy, and one which we'd suggest would be an excellent complementary one, would be to run the ^{15}N HMBC experiment. If alkylation were to have taken place on the central nitrogen (9), then the H12 protons should show correlation to three different nitrogens, but alkylation at either N8 or N10 could only result in two correlations. Again, a sound strategy but consider the likely resultant chemical shifts of the nitrogens of the triazole ring. For example, in the isomer shown, while N9 will have a very different shift from N8 and N10, there would be a danger that N8 and N10 could be accidentally equivalent, once again leading you astray as there would only be two correlations visible. It would be best to run both the HMBC and the ^{15}N HMBC experiments and to look for commonality of information between them. Danger lies everywhere so if you are forced by whatever circumstances to put all your eggs in one basket, it's best to make sure it has a strong handle!

Now consider the close analogue shown below...

In this case, another risk presents itself, in addition to the possibility of accidental equivalence discussed above – one you need to be on top of. Tautomerism of the imidazole NH might very well broaden the carbon signals of the imidazole and this broadening could extend as far as the key quaternary carbons of the triazole to the extent that these may be too broad to show the crucial 3-bond correlations in the HMBC experiment that we would be relying on to deduce the site of benzylation. (Note that tautomeric broadening would be likely because of asymmetry in the imidazole.) Mitigation of this risk would entail a careful examination of 1-D data to assess extent of signal broadening and possibly re-running the experiments at higher temperature to combat the signal broadening by speeding the rate of tautomeric inter-conversion.

Now we can have a look at a few examples and see how some of these ideas work in practice. In the following pages, we'll try to un-pick the entire thought processes that lead to the successful resolution of a few problems.

Let's presuppose that we have purchased a compound from an unfamiliar supplier and that it is extremely important that its structure is correct as we need to embark on a long and costly synthesis using it as a starting material. What do we need to do to satisfy ourselves that our efforts won't be wasted? The proposed structure of the compound is shown below:

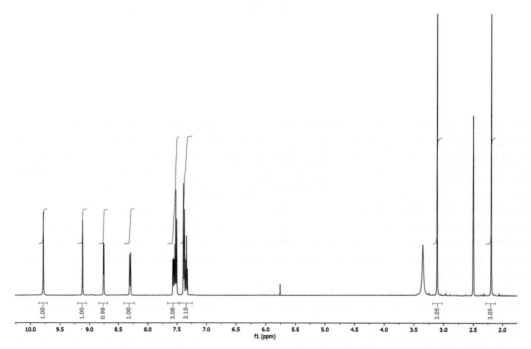

In Chapter 5 we certainly touched on some key aspects of approaching spectral data. The point was made that one of the most daunting features of an NMR spectrum can be the sheer quantity of information that confronts you – and this was in relation to the basic 1-D proton spectrum. If you collect a full set of 2-D spectra as well, this information-overload gets a whole lot worse, so knowing what you're looking for and where to get it from, becomes even more important! Perhaps somewhat surprisingly, the availability of the 2-D techniques, does not render the 1-D proton spectrum obsolete. Far from it! They give it a new significance because if the humble proton spectrum fails to give you the information you need, it can usually provide the next best thing – a 'road map' of where you need to be going with your problem. So, where do we start with our compound? With the 1-D proton spectrum (Spectra 17.1 and 17.2)! Note that all spectra for this example were acquired on a 500 MHz instrument.

Spectrum 17.1   Proton spectrum of purchased compound.

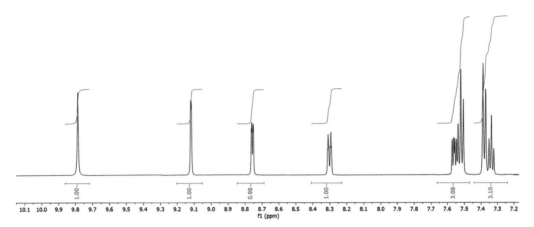

Spectrum 17.2   (Expansion).

The first observations will certainly be positive. The compound looks nice and pure with only a trace of what looks like dichloromethane to be seen as far as impurities are concerned. The spectrum is a relatively simple one but a look at an expanded plot of the aromatic region would certainly be worthwhile. This confirms, at a glance, the mono-substituted pyridine ring with shifts and couplings that are very satisfactory for the proposed structure. We can also see a mono-substituted benzene ring quite clearly, along with two methyl singlets in positions that are entirely in-keeping with the proposed structure. Returning briefly to the benzene ring, the chemical shift of H7/11 would be difficult for us to estimate with any accuracy, as this would be very much dependent on the orientation of the benzene ring with respect to the rest of the molecule. However, the shift of the H7/11 protons (7.39 ppm) indicates that the O17 is not deshielding these protons to any great extent and that therefore the two rings are considerably out-of-plane with each other.

A useful tool, when looking at spectra in detail, is to ask yourself the question, 'Could any other related or similar compound give a spectrum similar to the one I'm looking at and if so, how can I be sure that I've got the structure I need?' It's a case of playing 'devil's advocate' with yourself. It can be a useful exercise! In this case, for example, how can we be sure of the arrangement of the components of the five-membered ring? What if the N-Me and the N-AR groups were swapped around? How could we be sure of their relative positions? Firstly, check out the HSQC, just to be certain that the N-Me and C-Me groups have plausible shifts. (See Spectrum 17.3.)

[13]C shifts of 36.4 ppm and 11.5 ppm are entirely in line with expectations, so no problem here. So this is where we need a 'magic bullet' to kill off any possible doubts that we might have, regarding the authenticity of this sample. Look at the proposed structure again. Look for a key supporting feature which you really must see to satisfy yourself that all is well…

If the C and N methyls are next to each other as in our proposed structure, then there should be an NOE between the H12 and H13 protons, so an NOE or ROESY experiment would be a good idea. An alternative approach would be an HMBC experiment which must show a common correlation with the quaternary carbon, C3. (See Spectrum 17.4.)

You can see the key correlations to C3 (153.8 ppm), quite clearly; the two strong correlations from both methyls (H12 and 13) as well as a weaker one from NH14. (Note that cis correlations like this are often weak.) Another important observation is the identification of both the carbonyl carbons. C15 shows correlations to NH(14), as well as to H23, as expected. (There is a correlation to H19 as well, but it is too weak to show on the vertical scale plotted above.) C5, at 162.3 ppm on the other hand, is in a position where it may only show a weak, cis correlations

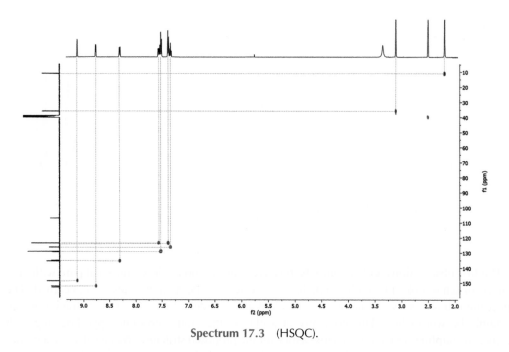

Spectrum 17.3 (HSQC).

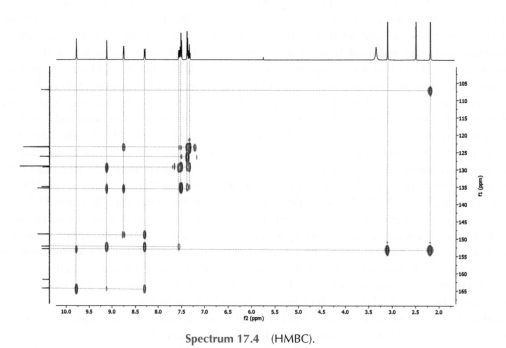

Spectrum 17.4 (HMBC).

to NH14, but this is to be expected from the proposed structure and is, therefore, a worthwhile observation.

So the key observations have been made and the structure fits the spectral data. A full assignment of all carbons and a record of all observed correlations from the HSQC and HMBC spectra would be recommended for reference but the case for the structure has been made. To summarise, we have:

1. Established presence of a mono-substituted pyridine, substituted with a carbonyl function (C15) at C18, confirmed by HMBC from H23 and 19.
2. Confirmed secondary amide (correlation from H14 to carbonyl C15).
3. Confirmed that both C-Me and N-Me are next to each other on the five-membered ring. Note also that the C-Me is next to the junction point, C4, as indicated by HMBC correlation between H13 and C4 (107.5 ppm).
4. Established presence of a mono-substituted benzene ring. Note that there are no correlations from any of the aryl protons to any carbons that aren't part of the aryl ring which underpins the proposed structure with the benzene ring attached to a nitrogen.

By using the data in this way, we have effectively tied the whole structure together leaving no room for uncertainty. A 2-D ROESY and a ^{15}N HMBC would have been 'nice-to-haves' but would not be regarded as essential for resolution of this structure, though a supporting mass ion would be highly recommended.

In our second example (also acquired at 500 MHz), the proposed structure for the compound under scrutiny is as follows:

Once again, we look at the 1-D proton spectrum first. (See Spectrum 17.5.)

At first sight, the spectrum appears to be very much in order. We immediately note the key features such as the broad exchangeable signal at approx. 10.5 ppm which is an excellent candidate for the phenolic–OH. (Note that the integration is poor for a single proton but the integration of exchangeable signals should always be viewed with a degree of scepticism as they can deviate from the expected value due to association with water or because their breadth can make their integration less reliable.) We also note an aryl AA'BB' system with protons at the expected chemical shifts for the proposed substituents, four further aryl protons which look perfectly believable for a 1,2 disubstituted aryl system – see Spectrum 17.6 (expansion) for an expansion of the aryl region and, of course, an equally satisfactory ethyl group exhibiting textbook chemical shifts.

So on the basis of this straightforward 1-D proton spectrum which shows no overlapping or confusing signals and given that we have a satisfactory mass ion, everything looks just fine. So should we accept that the structure fits the spectrum and proceed? Maybe – but perhaps using the 'devil's advocate tool' outlined in the last example might be worthwhile... So what other

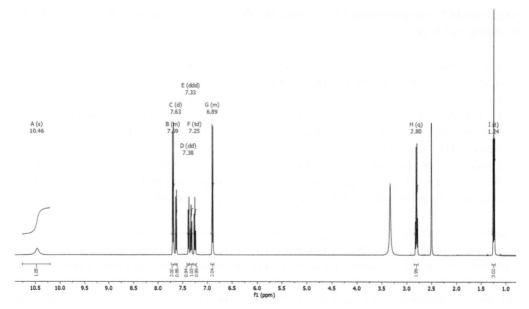

Spectrum 17.5 Proton spectrum of second example.

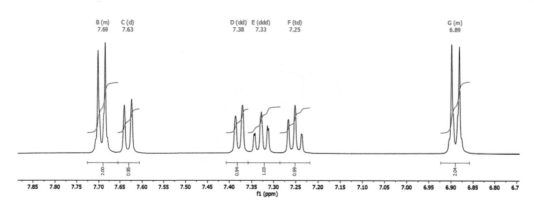

Spectrum 17.6 (Expansion).

plausible structure might give a very similar set of signals to the ones we see in this spectrum? With so much of the spectrum 'nailed down', we have little room for manoeuvre…but what if the substituents at C2 and C3 were to be reversed? Would we be any the wiser with only a 1-D proton to look at? The answer has to be a definite 'No!' and we really need more data to probe this structure in order to have full confidence in it.

The first consideration must be one of how best to discriminate between the two potential structures. In the case of the proposed structure, an NOE between H12 and H9 would be expected and if such an NOE were to be observed, the proposed structure could be considered pretty much confirmed. However, it is worth remembering that the absence of an expected NOE should not be considered as definitive proof of a structure, as there might be other over-arching reasons for the absence. If we had both regioisomers to work with and a clear NOE from the ethyl CH$_2$ protons was to be observed in one of them and not in the other, then the case could be considered closed.

For a safer resolution of the structure, given that we have only a single compound, we need to consider another approach. Let us consider the two structures again…

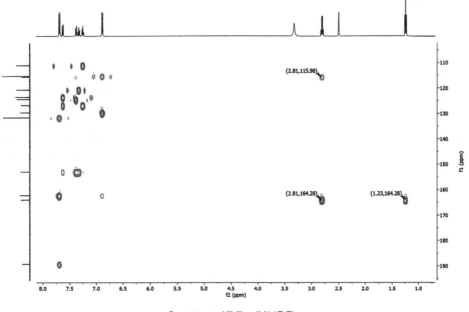

Proposed structure   Possible alternative structure

In the case of the proposed structure, the H12 protons would be expected to show correlations to C2, C3 and C4 whereas in the possible alternative structure, the corresponding methylene protons (H10) could only show two correlations, to C2 and C3. Furthermore, the chemical shift of the quaternary carbon to which the ethyl group is attached will be highly diagnostic and can be unambiguously assigned as, in both cases, it will be the only aryl carbon which can correlate to the methyl protons. The relevant part of the HMBC spectrum is shown in Spectrum 17.7.

The two key features discussed above are immediately apparent. Firstly, the methyl protons show a correlation to a carbon at a shift of 164.5 ppm. Secondly, the methylene protons show only two correlations; to 164.5 ppm as expected (a 2-bond correlation) and a 3-bond correlation to a carbon at 116.2 ppm. This evidence supports the possible alternative structure and is extremely compelling and the case can be considered closed. (This second correlation is of

Spectrum 17.7   (HMBC).

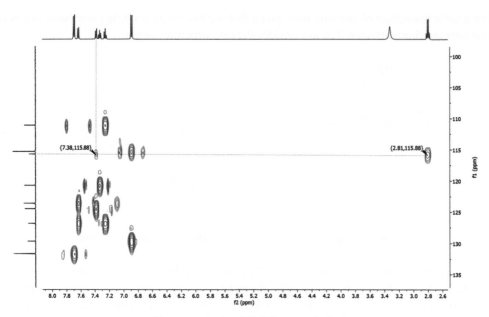

Spectrum 17.8   (HMBC expansion).

course to C3.) As a general observation, it is always a good idea to further underpin your deductions by using a good ^{13}C prediction package, such as the ACD software that we have mentioned before. This can give a further level of security and can flag any outlandish shifts that you might encounter. In this example, ^{13}C predictions of both proposed and possible alternative structures further highlight the differences between them.

Note that the resolution of problems like this is driven by entirely logical processes. Nothing can be left to chance or assumed. Conflicting evidence cannot be left unresolved. All the pieces of evidence must fit together as a solution evolves.

As a final footnote to this example, a further expansion of the HMBC (Spectrum 17.8) reveals a common correlation between the C3 carbon we've assigned by correlation to H10 and an aromatic proton showing a single large ortho coupling. This proton can only be H9. Note that the correlation to C3 is relatively weak as 3-bond correlations go because of the 'cis' relationship between the proton and carbon in question. This is quite typical and indeed, characteristic for this type of geometry. The positive identification of H9 automatically provides an assignment for H6 and for H7 and H8 via a COSY spectrum. From there, it would be a relatively easy step to provide a complete, unequivocal assignment in both proton and carbon domains. This is the 'gold standard'. Everything would be assigned and inter-related, leaving quite literally, no room for doubt.

In our third and final example, we'll have a look at a slightly more complex problem. In this case, the proposed structure for the compound in the spotlight is shown below.

The spectra relating to this example were all acquired on a 600 MHz instrument. Take a look at the 1-D proton spectrum (Spectrum 17.9).

Straight away, we can see that all is not as it should be. The broad signal at about 11.5 ppm shouldn't be there at all and given the shift and width of the signal, it must clearly be some sort of exchangeable proton. The second feature that is noteworthy is the number of broad signals between 2.5 and 3.7 ppm. Within the context of the proposed structure, the only broad signal expected in this region of the spectrum would be the water in the DMSO. Clearly, we need to look deeper… Consider the expansions of the proton spectrum (Spectra 17.10, 17.11 and 17.13).

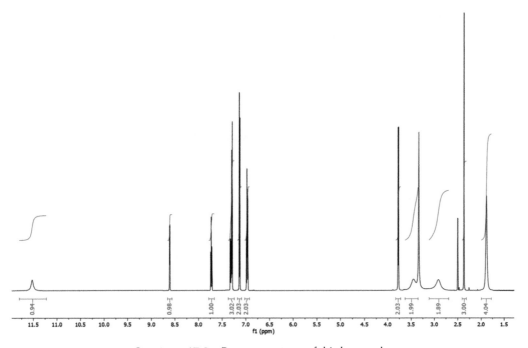

Spectrum 17.9 Proton spectrum of third example.

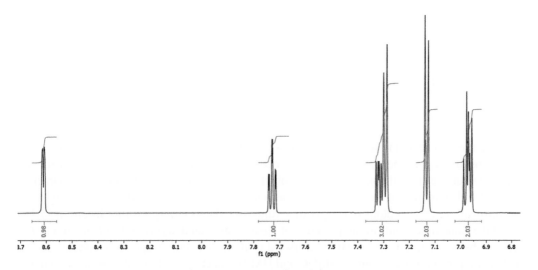

Spectrum 17.10 (Expansion 1).

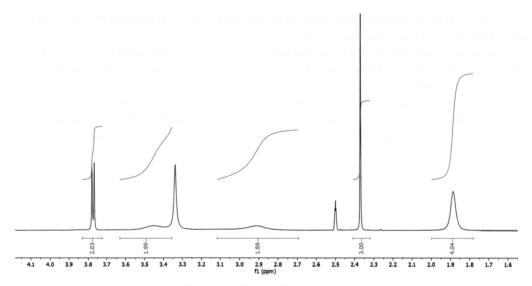

Spectrum 17.11   (Expansion 2).

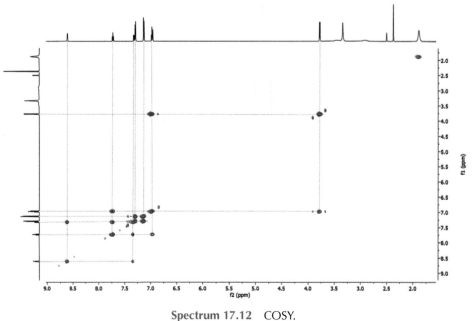

Spectrum 17.12   COSY.

Taking the aryl region first, it is clear that we have excellent evidence for the presence of a mono-substituted pyridine ring which is attached to something at the correct position for the proposed structure (note the narrow broad doublet at about 8.6 ppm which shows the characteristic shift and coupling which would be as expected for H11). We can use this as a starting point to confirm all the other protons of the pyridine ring using the COSY spectrum (Spectrum 17.12) and, at the same time, confirm the presence of a para-substituted aryl ring. Confirmation of the methyl substituent on this ring is straightforward from the 1-D proton spectrum and this can be easily underpinned further from the HSQC and HMBC spectra which we'll look at a bit later.

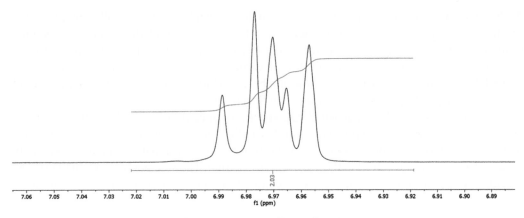

Spectrum 17.13   (Expansion 3).

Progress is being made but we now need to un-pick the two-proton signal at around 6.9–7.0 ppm. We know from the COSY spectrum that this signal must incorporate pyridine signal H14 and, keeping faith with the proposed structure for the moment, one other which can only be the alkene signal H8. This signal is worth an extra expansion all of its own (Spectrum 17.13).

Considering this signal in greater detail, it can be seen that it is composed of a triplet and a slightly broader doublet. Since the COSY clearly indicates that one of the signals in this multiplet must be that of pyridine proton H14, the alkene, H8 would have to be responsible for the triplet…? And now we have a serious problem. Given the proposed structure, the alkene proton cannot present as a triplet with such a large splitting.

At this point, we have demonstrated that the proposed structure cannot be correct, and our investigation must enter an completely new phase. Up until this point, our thinking has been entirely analytical but now, we must bring a certain measure of creativity to the problem. We need to consider what features in our structure might be responsible for bringing about the spectral features that we see and never lose sight of the fact that there is a *reason* for everything we see in an NMR spectrum. Our job is to understand these reasons – and to translate them into potentially viable structures.

If the alkene proton is a triplet, then, any heteroatoms aside, it must be situated next to two other protons. The COSY spectrum shows a coupling between this triplet and the doublet at 3.8 ppm. This implies that the groups on the alkene need to be re-arranged so that the alkene CH is next to a CH_2 giving rise to one of the structure types shown below*. (Note that we haven't yet resolved the issue of the broad alkyl signals.)

---

*Some of you might have considered the possibility of using the HMBC spectrum to establish a common correlation between H3 on the pyridine and H16/20 on the toluene ring with alkene carbon C7. While this would be an excellent line of investigation in principle, in this particular example the overlap of H3 with H8 hampers particular observation. With both H3 and H8 in a position to correlate to C7, it would be unwise to attempt to draw a conclusion from any observed correlation.

At this stage, we need to return to the broad signals that can be seen in the alkyl region of the spectrum (see Spectrum 17.11). Can they be explained in terms of the proposed structure? Could there be something like an amide group in there causing the broadening by restricted rotation? This would give a carbon shift for the carbonyl at around 170 ppm so a quick check of the 1-D ^{13}C and/or the HMBC spectra would be a good idea at this stage, though this should of course be flagged-up by mass spectrum. See Spectrum 17.14.

A close inspection of the expanded HSQC spectrum (Spectrum 17.15) shows that the broad proton signals at 2.90 and 3.45 ppm both correlate to the same carbon. Note that the carbon shift of this carbon and the carbon associated with the proton doublet at about 3.75 ppm are extremely close and are only just resolved in the 1-D carbon spectrum. They are not resolved in the HSQC spectrum as digital resolution in the 2-D spectra can never be as good for obvious reasons.

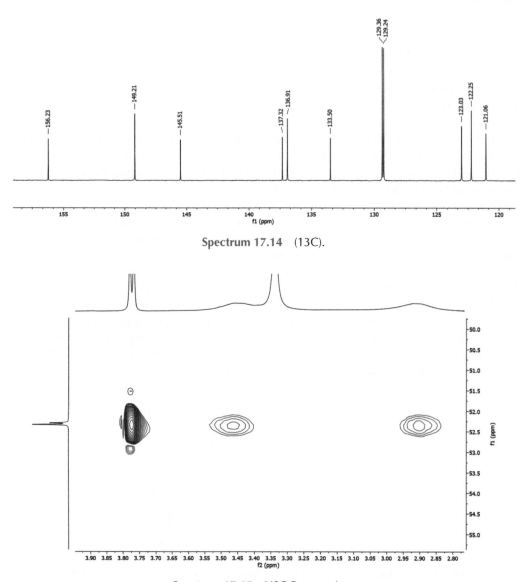

Spectrum 17.14   (13C).

Spectrum 17.15   HSQC expansion.

The explanation for this feature of the spectrum is that the compound is in fact a salt with the nitrogen of the pyrrolidine ring being protonated. Inversion of this nitrogen is relatively slow on the NMR timescale which gives rise to broad signals for the protons on each face of the pyrrolidine ring as they are on different steric environments. (This phenomenon has been discussed in detail in Section 6.8.)

The presence of an acid is further confirmed by the broad signal seen in the proton spectrum at about 11.5 ppm. Since the way is now clear to accept the pyrrolidine side chain, we can continue to build on what we know and consider the two candidate molecules shown below. These can be unambiguously assigned by an NOE-type experiment. In this case, we have opted for the 1-D ROESY and have targeted the H9 protons as they are not close enough to any other signal to cause any potentially confusing secondary enhancements. Note that the same cannot be said of the initially attractive H8 alkene proton on account of its overlap with H3 which is one of the key protons which might be expected to show an enhancement in the case of the first isomer. The ROESY spectrum can be seen in Spectrum 17.16. (Note also, the NOE between H9 and the broad H11/14 protons.)

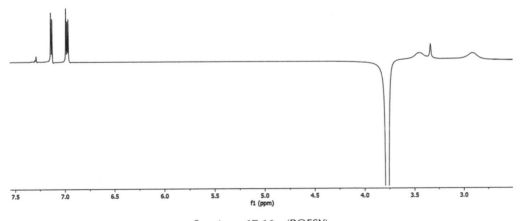

Spectrum 17.16 (ROESY).

What we are looking for, of course will be the enhancement of either H16/20 in the case of the first isomer or H3 in the case of the second isomer. As we can see, there is a clear enhancement of one half of the aryl AA'BB' system at about 7.15 ppm indicating that the sample must be of the first isomer. Note that the enhancement at 6.98 ppm is that of alkene proton, H8. Its enhanced signal is not that of a clear triplet because it is spin-coupled to the protons that have been irradiated and this invariably leads to distortions of this type. So, there we have it. All the relevant information has been considered and all issues addressed so our confidence in the structure can be high.

Hopefully, these examples will have helped illustrate the level of scrutiny required and the sort of inter-related arguments that you must apply to your data in order to bring your problems to a successful conclusion.

A few final words of advice… Keep an open mind. Concentrate on the spectra rather than the chemistry and *never* back yourself into a corner of the type '…this reaction can *only* give 'A' or 'B''. We've heard this so many times over the years, only to find that it gave 'C' or 'D'.

Appendix A

NMR Interpretation Flow Chart

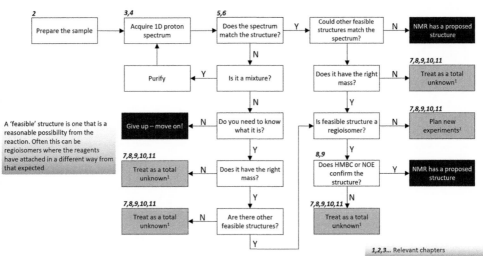

Don't forget! NMR on its own cannot prove a structure.

Figure A.1   Useful thought processes in tackling NMR problems.

Essential Practical NMR for Organic Chemistry, Second Edition. S.A. Richards and J.C. Hollerton.
© 2023 John Wiley & Sons Ltd. Published 2023 by John Wiley & Sons Ltd.

Glossary

Note: This glossary is by no means exhaustive but it hopefully contains most of the more important terms you will come across in a typical 'NMR environment'. Some of the entries may not even have featured in the text itself. While every effort has been made to make the entries scientifically valid, please note that it is sometimes difficult to condense a highly complex topic into a pithy three-line explanation, so some of the definitions are sketchy to say the least!

Acquisition Process of collection of NMR data.

Adiabatic pulse A type of pulse employing a frequency sweep during the pulse. This type of pulse is particularly efficient for *broadband decoupling* over large sweep widths.

Aliased signals Signals that fall outside the *spectral window* (i.e. those that fail to meet the *Nyquist condition*). Such signals still appear in the spectrum but at the wrong frequency because they become 'folded' back into the spectrum and are characterised by being out of phase with respect to the other signals.

Anisotropy Non-uniform distribution of electrons about a group which can lead to non-uniform localised magnetic fields within a molecule. The phenomenon leads to unexpected chemical shifts – particularly in ^1H NMR – in molecules where steric constraints are present.

Apodization The use of various mathematical functions which when applied to an FID, yield improvements in the resultant spectrum. These include *exponential multiplication* and *Gaussian multiplication*.

Atropisomers Stereoisomers arising from hindered rotation about a single bond where the energy barrier is high enough to allow isolation of the different conformers.

Bloch-Siegert shift A shift in resonant frequency of a signal which is in close proximity to a secondary applied r.f. The effect forces signals away from the applied r.f. and is only ever noticeable in homonuclear decoupling experiments where the applied r.f. and the observed signal can be very close.

Boltzmann Distribution The ratio of nuclei which exist in the *ground state* to those in the *excited state* for a sample introduced into a magnetic field – prior to any r.f. pulsing. This varies with probe temperature but primarily with magnet field strength.

Essential Practical NMR for Organic Chemistry, Second Edition. S.A. Richards and J.C. Hollerton.
© 2023 John Wiley & Sons Ltd. Published 2023 by John Wiley & Sons Ltd.

Broadband decoupling Decoupling applied across a wide range of frequencies, e.g. the decoupling of all proton signals during the acquisition of 1-D ^{13}C spectra.

CAMELSPIN Cross-relaxation Appropriate for Minimolecules Emulated by Locked spins. Now known as *ROESY*.

Chemical shift Position of resonance in an NMR spectrum for any signal relative to a reference standard.

Chiral centre An atom in a molecule (usually but not exclusively carbon) which is bound to four different atoms or groups such that the mirror image of the whole molecule is not superimposable on the molecule itself. A chiral centre in a molecule implies the possibility of the isolation of two distinct forms of the compound which are known as *enantiomers*.

Chirality Properties conferred by the presence of one or more *chiral centres*.

Composite pulses Use of a series of pulses of varying duration and phase in place of a single pulse. Such systems, when used in the pulse sequences of many modern NMR techniques, give improved performance as they are more tolerant to r.f. inhomogeneity.

Configuration The arrangement of atoms and bonds in a molecule. The configuration of a molecule can be changed by breaking and re-forming bonds to yield different *regioisomer*.

Conformation The shape a molecule adopts by the rotation and deformation (but *not* the breaking and re-forming) of its bonds.

Continuous Wave (CW) Technology used initially in the acquisition of NMR data. The radio-frequency or the magnetic field was swept and nuclei of different *chemical shift* were brought to resonance sequentially.

COSY COrrelative SpectroscopY. Homonuclear (normally ^{1}H) 2-D spectroscopic technique which relates nuclei to each other by spin coupling.

Coupling The interaction between nuclei in close proximity which results in splitting of the observed signals due to the alignment of the neighbouring nuclei with respect to the magnetic field. Also referred to as *spin coupling*.

Coupling constant The separation between lines of a coupled signal measured in Hz.

CPMG pulse sequence Carr–Purcell–Meiboom–Gill pulse sequence. A pulse sequence used for removing broad signals from a spectrum by multiple de-focusing and re-focusing pulses.

Cryoprobe *Probe* offering greatly enhanced sensitivity by the reduction of thermal electronic noise achieved by maintaining probe electronics at or near liquid helium temperature.

Cryoshims Rough (*superconducting*) shim coils that are built into superconducting magnets and adjusted at installation of the instrument.

Decoupling The *saturation* of a particular signal or signals in order to remove spin coupling from those signals. Also referred to as *spin decoupling*.

DEPT Distortionless Enhancement by Spectrum Editing. A useful 1-D technique which differentiates methyl and methine carbons from methylene and quaternary carbons.

Diastereoisomers *Stereoisomers* that are not *enantiomers*. Diastereoisomers are compounds that always contain at least two centres of chirality.

Diastereotopic proton/group A proton (or group) which if replaced by another hypothetical group (not already found in the molecule), would yield a pair of diastereoisomers.

Enantiomer A single form of an optically active compound. Optically active compounds usually (but not exclusively) contain one or more chiral centres. Enantiomers are defined by their ability to rotate the plane of a beam of polarised light one way or the other and these are referred to as either 'D' or 'L', or alternatively '+' or '−', depending on whether the polarised light is rotated to the right (**D**extro) or the left (**L**evo) .

Enantiotopic proton/group A proton (or group) which if replaced by another hypothetical group (not already found in the molecule) would yield a pair of *enantiomers*.

Epimers Diastereoisomers related to each other by the inversion of only one of their *chiral centres*.

Epimerisation Process of inter-conversion of one *epimer* to the other. The process is usually base-mediated as abstraction of a proton is often the first step in the process.

Excited state Condition where nuclei in a magnetic field have their own magnetic fields aligned so as to oppose the external magnet, i.e. N-N-S-S. Also known as the high-energy state.

Exponential multiplication The application of a mathematical function to an *FID* which has the effect of smoothing the peak shape. Signal/noise may be improved at the expense of resolution.

First-order spin systems Not very specific term used to describe spin systems where the difference in *chemical shift* between coupled signals is very large in comparison to the size of the *coupling*. In reality, there is no such thing as a completely first-order system as the chemical shift difference is never infinite. See *non-first-order spin system*.

Folded signals See *aliased signals*.

Fourier transformation Mathematical process of converting the interference *free induction decay* into a spectrum.

Free Induction Decay (FID) Interference pattern of decaying cosine waves collected by Fourier Transform spectrometers, stored digitally prior to *Fourier Transformation*.

Gated decoupling A method of *decoupling* in which the decoupling is switched on prior to acquisition and turned off during it.

Gradient field A linear magnetic field gradient, deliberately imposed on a sample in, for example, the *z*-axis in order to defocus the magnetisation. This allows other refocusing gradient pulses to be used to selectively observe desired transitions. Only possible with appropriate hardware. Gradient fields improve the quality of many 2-D techniques and where used, replace the need for *phase cycling*.

Gradient pulse The application of a *gradient field* for a discrete period of time. Also referred to as *pulsed field gradients (PFGs)*.

Gaussian multiplication The application of a mathematical function to an *FID* to improve resolution (sharpen lines) at the expense of signal/noise.

GOESY Gradient Overhauser Effect SpectroscopY. An early version of a 1-D *NOESY* making use of gradients.

Gradient shimming A system of *shimming* based on mapping the magnetic field inhomogeneity using field gradients and calculating the required shim coil adjustments required to achieve homogeneity.

Ground state Condition where nuclei in a magnetic field have their own magnetic fields aligned *with* that of the external magnet, i.e. N-S-N-S. Also known as the low-energy state.

Gyromagnetic ratio A measure of how strong the response of a nucleus is. The higher the value, the more inherently sensitive will be the nucleus. ^1H has the highest value. Also known as *Magnetogyric ratio*.

Hard pulse A pulse which is equally effective over the whole chemical shift range. See *soft pulse*.

HETCOR HETeronuclear CORrelation. Early method of acquiring 1-bond ^1H–^{13}C data. Not nearly as sensitive as *HMQC* and *HSQC* methods which have largely superseded it.

HMBC Heteronuclear Multiple Bond Correlation. A proton-detected, 2-D technique that correlates protons to carbons that are two and three bonds distant. Essentially, it is an HMQC that is tuned to detect smaller couplings of around 10 Hz.

HMQC Heteronuclear Multiple Quantum Correlation. A proton-detected, 2-D technique that correlates protons to the carbons they are directly attached to.

HOHAHA HOmonuclear HArtmann HAhn spectroscopy. See *TOCSY*.

HSQC Heteronuclear Single Quantum Correlation. As for *HMQC* but with improved resolution in the carbon dimension.

INADEQUATE Incredible Natural Abundance DoublE QUAntum Transfer Experiment. Two-dimensional technique showing ^{13}C–^{13}C coupling. It should be the 'Holy Grail' of NMR methods but is in fact of very limited use due to extreme insensitivity.

Indirect detection Method for the observation of an insensitive nucleus (e.g. ^{13}C) by the transfer of magnetisation from an abundant nucleus (e.g. ^{1}H). This method of detection offers great improvements in the sensitivity of proton–carbon correlated techniques.

Inverse geometry Term used to describe the construction of a probe that has the ^{1}H receiver coils as close to the sample as possible and the X-nucleus coils outside these ^{1}H coils. Such probes tend to give excellent sensitivity for ^{1}H spectra at the expense of X-nucleus sensitivity in 1-D techniques. They offer a lot of compensation in terms of sensitivity of *indirectly detected* experiments.

J-resolved spectroscopy Two-dimensional techniques, both homo- and heteronuclear, that aim to simplify interpretation by separating chemical shift and coupling into the two dimensions. Unfortunately prone to artefacts in closely coupled systems.

Laboratory frame model A means of visualising the processes taking place in an NMR experiment by observing these processes at a distance, i.e. with a static co-ordinate system. See *rotating frame model*.

Larmor frequency The exact frequency at which nuclear magnetic resonance occurs. At this frequency, the exciting frequency matches that of the precession of the axis of the spin of the nucleus about the applied magnetic field.

Larmor precession The motion describing the rotation of the axis of the spin of a nucleus in a magnetic field.

Linear prediction Method of enhancing resolution by artificially extending the *FID* using predicted values based on existing data from the *FID*.

Longitudinal relaxation (T_1) Recovery of magnetisation along the z-axis. The energy lost manifests itself as an infinitesimal rise in temperature of the solution. This used to be called *spin-lattice relaxation*, a term which originated from solid-state NMR.

Magic Angle Spinning (MAS) $54°44'$ (from the vertical). Spinning a sample at this, the so-called magic angle gives the best possible line shape as the broadening effects of chemical shift anisotropy and dipolar interactions are both minimised at this angle. Used in the study of molecules tethered to solid supports.

Meso compound A symmetrical compound containing two *chiral centres* configured so that the *chirality* of one of the centres is equal and opposite to the other. Such internal compensation means that these compounds have no overall effect on polarised light. e.g. meso tartaric acid.

Normal geometry Term used to describe the construction of a conventional dual/multi-channel probe. Since the X-nucleus is a far less sensitive nucleus than ^{1}H, a 'normal geometry' probe has the X-nucleus receiver coils as close to the sample as possible to minimise signal loss and the ^{1}H receiver coils outside the X-nucleus coils i.e. further from the sample. This design of probe is thus optimised for X-nucleus sensitivity at the expense of some ^{1}H sensitivity.

NOE Nuclear Overhauser Effect/Nuclear Overhauser Enhancement. Enhancement of the intensity of a signal via augmented relaxation of the nucleus to other nearby nuclei that are undergoing saturation.

NOE Nuclear Overhauser Experiment. Experiment designed to capitalise on the above. Such experiments (and related techniques, e.g. *NOESY* etc.) are extremely useful for solving stereochemical problems by spatially relating groups or atoms to each other.

NOESY Nuclear Overhauser Effect SpectroscopY. Two-dimensional technique that correlates nuclei to each other if there is any *NOE* between them.

Non-first-order spin system Spin system where the difference in chemical shift between coupled signals is comparable to the size of the coupling between them. These are characterised by heavy distortions of expected peak intensities and even the generation of extra unexpected lines.

Nyquist condition Sampling of all signals within an FID such that each is sampled at least twice per wavelength.

Phase The representation of an NMR signal with respect to the distribution of its intensity. We aim to produce a pure absorption spectrum (one where all the signal intensity is +ve).

Phase cycling The process of repeating a pulse sequence with identical acquisition parameters but with varying r.f. phase. This allows real NMR signals to add coherently while artefacts and unwanted NMR transitions cancel.

Phasing The process of correcting the *phase* of a spectrum (either manually or under automation).

Probe Region of the spectrometer where the sample is held during the acquisition of a spectrum. It contains the transmitter and receiver coils and gradient coils (if fitted).

Pulse A short burst of radio frequency used to bring about some nuclear spin transition.

Pulsed field gradients (PFGs) See *gradient pulse*.

Quadrature detection Preferred system of signal detection using two detection channels with reference signals offset by 90°.

Quadrupolar nuclei Those nuclei, which because of their *spin quantum number* (which is always >1/2), have asymmetric charge distribution and thus possess an electric quadrupole as well as a magnetic dipole. This feature of the nucleus provides an extremely efficient relaxation mechanism for the nuclei themselves and for their close neighbours. This can give rise to broader than expected signals.

Quadrupolar relaxation Rapid relaxation experienced by *quadrupolar nuclei*.

Racemate A 50/50 mixture of *enantiomers*.

Regiochemistry The chemistry of a molecule discussed in terms of the positional arrangement of its groups.

Regioisomers Isomeric compounds related to each other by the juxtaposition of functional groups.

Relaxation The process of nuclei losing absorbed energy after excitation. See *longitudinal relaxation* and *transverse relaxation*.

Relaxation time Time taken for *relaxation* to occur.

ROESY Rotating-frame Overhauser Effect SpectroscopY. A variation (one and two dimensional) on the Nuclear Overhauser Experiment (*NOE*). The techniques have the advantage of being applicable for all sizes of molecule. See *laboratory frame model*.

Rotating frame model A means of visualising the processes taking place in an NMR experiment by observing these processes as if you were riding on a disc describing the movement of the bulk magnetisation vector.

Saturation Irradiation of nuclei such that the slight excess of such nuclei naturally found in the *ground state* when a sample is introduced into a magnet, is equalised.

Shim coils Coils built into NMR magnets designed to improve the homogeneity of the magnetic field experienced by the sample. Two types of shims are used: *cryoshims* and room temperature shims. Normal shimming involves the use of the room temperature shims.

Shimming The process of adjusting current flowing through the room temperature *shim coils* in order to achieve optimal magnetic field homogeneity prior to the *acquisition* of NMR data. The process may be performed manually or under automation.

Soft pulse Pulse designed to bring about irradiation of only a selected region of a spectrum. See *hard pulse*.

Solvent suppression Suppression of a dominant and unwanted signal (usually a solvent) either directly by saturation or by use of a more subtle method such as the *WATERGATE* sequence.

Spectral window The range of frequencies observable in an NMR experiment.

Spin coupling See *coupling*.

Spin decoupling See *decoupling* and *broadband decoupling*.

Spin quantum number Number indicating the number of allowed orientations of a particular nucleus in a magnetic field. e.g. ^{1}H has an I value of ½, allowing for two possible orientations whereas ^{14}N has an I of 1, allowing three possible orientations.

Spin–lattice relaxation See *longitudinal relaxation*.

Spin lock A pulse sequence that keeps magnetisation in the transverse plane for a certain time.

Spin–spin relaxation See *transverse relaxation*.

Stereochemistry The chemistry of a molecule discussed in terms of its 3-D shape.

Stereoisomers Diastereoisomers related to each other by the inversion of any number of chiral centres.

Superconduction Conduction of electric current with zero resistance. This phenomenon occurs at liquid helium temperature and has made possible the construction of the very high powered magnets that we see in today's spectrometers.

TOCSY TOtal Correlation SpectroscopY. One and two-dimensional techniques that are analogous to *COSY* but which differ in that it shows couplings within specific spin systems.

Transverse relaxation (T_2) Relaxation by transfer of energy from one spin to another (as opposed to loss to the external environment as in *longitudinal relaxation*). This used to be referred to as *spin–spin relaxation*.

WATERGATE WATER suppression through GrAdient Tailored Excitation.

Zero filling Cosmetic improvement of a spectrum achieved by padding out the *FID* with zeros.

Index

Essential Practical NMR for Organic Chemistry, Second Edition. S.A. Richards and J.C. Hollerton.
© 2023 John Wiley & Sons Ltd. Published 2023 by John Wiley & Sons Ltd.

Printed and bound by CPI Group (UK) Ltd, Croydon, CR0 4YY

27/10/2024

14580304-0003